SENSATIONAL MOMENTS IN SCIENCE

KARL KRUSZELNICKI
SENSATIONAL MOMENTS IN SCIENCE

illustrations by GRAHAM and ANN BEATTY

an
ABC
BOOK

Published by ABC Books for the
AUSTRALIAN BROADCASTING CORPORATION
GPO Box 9994 Sydney NSW 2001

First published November 1995
Reprinted February 1996
Reprinted October 1997

National Library of Australia
Cataloguing-in-Publication entry

Kruszelnicki, Karl, 1948.
  Sensational moments in science.

  ISBN 0 7333 0456 7.

  1. Science – Popular works. I. Australian Broadcasting
  Corporation. II. Title. III. Title: Great moments in
  science (Radio program).
500

Designed by Geoff Morrison
Illustrated by Graham and Ann Beatty
Photography by Peter White, Karl Kruszelnicki
Edited by Margaret Bowman
Set in 11 / 15 pt ITC Century Book by Keyset, Sydney
Printed and bound in Australia by
Australian Print Group, Maryborough, Victoria

Dedication

Whenever the Amish make a quilt, they always deliberately incorporate a mistake into it, to show that only God is perfect. I assume that there will be some mistakes in this book, and I hereby take blame for all of them. The first people to point out each mistake will get a free copy of my next book. I would like to dedicate this book to the many people that made it possible. Stuart Neal helped restart the whole 'XXX Moments in Science' saga in 1990, when he suggested that I should write another book. He has had the patience and understanding to help me along. Caroline Pegram has combined the various skills of research assistant, paintbox operator, audio-visual desk mixer and personal assistant into an undescribable job description, and has worked at the 'coal face' on this book, pulling together information, turning electro-magnetic wave-forms into hard copy, and solidfying concepts. Dan Driscoll (Producer, ABC National Program Unit, NPU) has been a sensitive and insightful nuturer of these concepts, and has helped advance them further. Paul Vadasz, as head of the ABC NPU, provided the environment in which these stories first appeared as short radio stories called 'Great Moments in Science', once per week, all over Australia. Without the radio stories, there would not have been any book stories.

Alex Gabbard from Oak Ridge National Laboratory helped correct 'Nuclear Coal', and Ted Ankrum of NASA was always able to give a technical answer. Ann and Graham Beatty provided the wonderful illustrations. Finally, Mary Dobbie and our children helped polish these stories, at many stages in their development.

CONTENTS

CHAOS MAKES MONEY

'CHAOS THEORY' HAS mysteriously emerged as a catch phrase over the last 10 years. Even though everybody thought that Chaos theory itself was totally useless, people everywhere talked knowledgeably about it and said 'crazy' things such as: 'A butterfly flapping its wings in the Amazon can set off a thunderstorm near New York'. But now Chaos theory is beginning to gain respectability, and some 'renegade' physicists are even trying to use it to become fabulously wealthy!

Now according to the ancient Greek myths of creation, Chaos was the dark silent abyss, or unorganised void, from which everything came. Chaos came even before the gods. It gave birth to Night, and to Erebus (the home of death). In turn, Night and Erebus gave birth to Love, which generated Light and Day. Chaos then acted upon this universe, which was empty of matter, to give birth to the world we have today. According to those ancient Greeks, this world is made up of the solid Earth, the star-filled Heaven, and Eros (or Love).

In more modern times, KAOS took the form of the evil organisation that Maxwell Smart battled against! This version is a much less romantic one than that of the Greeks.

One of the strengths of modern science is that it can often work out why things happen, and can often make accurate predictions. The French mathematician Pierre Simon de Laplace obviously held this belief firmly when he said in 1776:

Chaos in the Solar System

In the good old days before Chaos theory, everybody thought that if you had simple laws (or a few objects), you would get simple results. The solar system seemed pretty simple – there was just one central Sun and a bunch of only nine planets. The laws governing the movement of the planets looked pretty simple too – they were Kepler's three simple Laws of Planetary Motion. Applying these simple laws to our simple solar system should therefore result in a perfectly predictable and very simple system (a group of bodies all subject to the same forces). But there was a big surprise in store!

When Gerald Sussman (computer scientist) and Jack Wisdom (astronomer) from the Massachusetts Institute of Technology actually did an experiment based on these laws and plugged the velocities and starting positions of the planets into a big fat computer, their predictions after four million years were crazy. Even the tiniest variation in the initial position or the initial velocities of any of the planets led to enormous changes in the predicted positions of the planets – for example, Mars being close to Earth rather than far away.

In this case, simple and rigid laws led to chaos. So one thing that Chaos theory says is that you can get chaotic behaviour in a simple system.

'The present state of the system of nature is evidently a consequence of what it was in the preceding moment, and if we conceive of an intelligence which at a given instant comprehends all the relations of the entities of this universe, it could state the respective positions, motions and general affects of all these entities at any time in the past or future'. In other words, he was saying that if we knew exactly the location and velocity of every particle in the universe, we could predict the future of the universe. But Laplace was wrong.

Although modern Chaos theory is still in its early stages, it does tell us a few things. It says that in the long term, it is not possible to predict things exactly. It also says that in the short term, it is sometimes possible to make pretty good predictions.

To the Chaos scientists, 'chaos' is a behaviour or event so complicated and irregular that it appears to follow no natural laws, and so it appears random. The formation of clouds seems chaotic. A chaotic system is one that is extremely sensitive to tiny variations in the initial conditions. The weather is a good example of a chaotic system – where a very small change at one time and place, can lead to a major change later on.

That's the basis of the butterfly effect – the famous fictional example of the butterfly beating its wings in the Amazon. Five days later, so the story goes, a storm will hit New York. But if the butterfly had not flapped its wings, the storm would have hit Boston! Of course, we certainly can't prove this particular case by doing the experiment and seeing what happens if the butterfly does, or does not, flap its wings – we would need two planets that were identical except for the flapping of this single butterfly! But it is an example of how some results are very sensitive to the starting conditions.

An American meteorologist, Edward Lorenz, first noticed this butterfly effect in the early 1960s. He was running a computer simulation of the weather to look at the movement of heat in the Earth's atmosphere. He had done one run, and decided to repeat it. He plugged in what he thought were the same starting conditions and began the computer run. The results he got from his new run were soon very different from those of his previous run. He could not understand why. Then he noticed that he had actually plugged in numbers that were different, but only by 0.1 per cent. He realised that the weather was a system that was very sensitive to the starting conditions – and therefore, very hard to predict in the long term.

In the short term, however, it is a different matter. Consider a rapidly flowing river that has a few boulders sticking out of the water. Every now and then, a tiny whirlpool will form near a boulder, break away, and rush downstream. We cannot predict when the whirlpool will form, nor can we predict what it will do half an hour after it forms. But we can predict that in the five seconds after it forms and starts to move, it will usually go downstream.

There is a similar rhythm in the stock market, as various influences come and go. Information is a wave that sweeps around our world, and in its trail are little whirlpools and eddies – and these can be used to make money. But the wave is a very complex one. It is shaped by forces we know about (supply and demand, mass psychology etc) and forces we don't know about. As an example of an unknown force, consider the old folk wisdom that says that the prices on the

Chaos in Behaviour

In the good old days before Chaos theory, everybody thought that if you had complicated laws (or lots of objects), you would get complicated results. In fact, they thought that the only way you could get complicated results was if you started with complicated laws (or lots of objects).

Suppose you looked at something that was really complicated, like human behaviour. Everybody thought it was like this because either you had a whole bunch of complicated laws, or you were dealing with enormous numbers of objects (like the billions of nerve cells inside each human brain).

But one robot scientist is getting surprisingly 'intelligent' and complicated behaviour from robots by using ridiculously simple laws. One such law might be: 'If you are low on energy, get your battery recharged'. Another law might be: 'If something attacks you, run away'. He is trying to build a humanoid robot from the 'bottom up'. Each section (arm, hand, leg, foot etc) will be run by different, and simple, laws. Maybe his approach will achieve success before the other roboticists, who are following a 'top-down' approach, and are programming robots with very complex laws.

Simplicity/ Complexity

Before the 16th century, people could make sense of the movement of the planets and the stars by explaining them as the whims of powerful gods. In the 16th century, the German astronomer Johannes Kepler reduced the motion of the planets to the three simple laws that guided them along elliptical orbits. Isaac Newton followed on from this work to discover a law of gravitation that applied to every object in the universe. This led to some people thinking of a predictable clockwork-type universe.

So the simple geometrical, elliptical shape of a planet's orbit was thought to be due to the law of gravity being so simple. The complexity of DNA, on the other hand, was thought to be a direct result of the enormous number of ways in which its atoms could be arranged.

stock market rise and fall with the hemlines of skirts!

Back in the 1970s in Santa Cruz in California, Doyne Farmer, with a bunch of other physicists, set up a 'science' commune called the 'Chaos Cabal'. They were obsessed by supposedly unpredictable events like dripping taps, weather – and roulette wheels. So they bought a second-hand roulette wheel from Las Vegas, to see if it was perfectly random or whether they could see anything predictable about it.

Now a roulette wheel is not completely random – nothing made by humans is perfect. The Amish deliberately put a mistake in every quilt that they make as a reminder that only God is perfect. The roulette wheel won't be set up perfectly level in the casino – it'll be a little bit heavier on one side than on the other side; the bearing holding the spinning wheel won't be perfectly even; and so on. So when you add in the factor that the croupier probably has a favourite way of throwing the ball (instead of throwing it differently each time), you can see that the combination of a roulette wheel and a croupier is definitely not a perfectly random system.

The unconventional Santa Cruz scientists spent weeks spinning the wheel and throwing the ball. Sure enough, they found 'islands of predictability' buried deep in the physics of the roulette wheel. These predictions were not very exact, however. They could not predict that the ball would land *on* 26 black, but they could say that it had a higher probability of landing in that one-eighth of the wheel that was centred *around* 26 black. One-eighth may not sound very good, but it's better than one-quarter, and a lot better than one-half of the roulette wheel. Knowing that the ball would land in one specific eighth of the wheel gave them a definite advantage – even if only a small one.

Now back in 1977, which was four years before the first usable personal computers appeared, most computers were enormous machines that filled three or four climate-controlled rooms. But in that same year these eccentric physicists from the Chaos Cabal built the first wearable computers – three thin slabs of electronics, with solenoid-

Past Predictions – Not So Accurate

operated buttons, that they wore inside their shoes! They used their toes to tap in data and commands. A small home computer today would have a RAM (Random Access Memory) of about 4000 kilobytes – their magic shoe computer had a total capacity of just 4 kilobytes!

In December 1977, they drove to Reno, Nevada, in their beat-up van. In the casino it was usually Farmer who would wear two of these magic computer shoe inserts – one in each shoe, joined by a cable running inside his trousers. Using his toes, he would program the two parts of his shoe computer with information such as where the ball landed on the wheel when it was thrown in, which cup it eventually landed in, and so on. It would take some time to feed in all the relevant information about one particular croupier operating one particular roulette wheel. (If that croupier left the table, they would of course have to start all over again.)

Now there were some 15 seconds between when the ball hit the roulette wheel and when it finished its chaotic run on the wheel by finally landing in a cup – but the tiny wearable shoe computers would work out the answer thousands of times faster than that.

Farmer's two shoe computers would talk to each other, work out an answer and then tell him where the ball was most likely to land – by tapping on his right big toe. Farmer would then use his left big toe to transmit that information, via a radio link, to the third computer shoe worn by a partner – it was important that the casino operators had no idea he and his partner were in collusion with each other. (In fact, the state of Nevada made such prediction systems illegal in 1985.) His partner would place the bets over a one-eighth section of the wheel, and more often than not, they would make money. In fact, they made enough money to convince them to go for the big one – the stock market.

If you look at the price of any particular stock, you will see that it goes up and down as the days go by. If you just guess at random whether a stock will go up or down the next day, you've got a 50 per cent chance of being right. But suppose

Chaos and Dynamics

Chaos began in the field of dynamics, which studies how systems (a group of bodies that are all subject to the same forces) change over time. It turns out that a chaotic system responds much more rapidly than a non-chaotic system to an outside stimulus.

Some modern jet fighters are made deliberately unstable so that they can change direction more quickly than would a stable jet. If the computers controlling the jet all fail, then the fighter falls out of the sky, because the human pilot does not have reactions quick enough to control the jet.

Tennis players waiting to receive a serve often dance erratically from one foot to another, making rapid tiny movements all the time, and continually changing their direction. This movement makes them unstable, and they would fall over if they were not continually making corrections to their movements. They say that this lets them respond rapidly to a ball coming very quickly from any direction.

The heart, when it is at rest, seems to have a regular rhythm. But if you measure it accurately, you'll find that the time from one heart-beat to the next always changes by tiny amounts. Maybe a 'chaotic' heart can respond more rapidly to sudden stress.

Past Predictions – Accurate

One of the most accurate predictions ever made came true at 8.56 pm Eastern Pacific Time, on Thursday, 24 August 1989. The *Voyager 2* spacecraft zipped past Neptune, just 4905 kilometres above the cloud tops. It was within one minute, and a few kilometres, of the time and position that its controllers had predicted it would be in, some three-and-a-half years earlier! Their prediction was 99.999997 per cent accurate!

The ancient Greeks built a calculator that could predict the timing of eclipses to an accuracy of 99.9975 per cent! This calculator contains 32 bronze gears, and a differential gear. It was recovered in 1901 from an ancient Greek ship that had sunk off the Aegean island of Antikythera. At that stage it looked like a lump of metal that sea animals had grown over. But an analysis with X-rays in 1972 found the metal gears inside. The Antikythera device was built during the 1st century BC.

The ancient Egyptians could also make accurate predictions. They needed to know when the Nile River would flood, dumping its fertile silt on the farmers' land. They knew that this would usually happen after the first appearance of the star Sirius in the night sky.

Those ancient astronomers were the first scientists. The motto of the philosopher is 'I think, therefore I am'. But the motto of these astronomers might have been 'I think, therefore I get paid'!

you're a big bank, and suppose your team of tame renegade physicists increases those odds to 55 per cent. Then you've got a 5 per cent leverage. Suppose you can bet $20 million. Five per cent of $20 million is $1 million, and so at the end of one day's trading you've made $1 million profit – which is not bad for only one day's work.

Farmer and his friends have since teamed up with some very big investors and formed the Prediction Company, to play the stock market. Other investors have their tame scientists too. Citibank has had the British mathematician Alan Colin on the payroll since 1990, while the London Midland Bank has a team of eight scientists.

It was handy for Farmer and his friend Packard that they had started out in physics. It turns out that the equations used to deal with the stock market are the same equations used to deal with heat flow. It's no coincidence that over the last five years, about half of the PhD graduates in theoretical physics from both Harvard and Stanford universities in the USA have left physics and are now working in finance.

This whole concept of getting wealthy by working the stock market seems a little tacky, though, when you compare it to what used to be called 'honest labour', which meant that you actually made a product, like a railway wheel, or a house.

Chaos theory seems to back up the beliefs expressed by Carl Jung, the Swiss psychiatrist who wrote about the collective unconscious, when he said, 'In all chaos there is a cosmos, in all disorder a secret order'. Or maybe Secret Agent 86, Maxwell Smart, was on the right track when he battled the agents of his arch-enemy, KAOS, with his shoe-phone?

References
Scientific American, December 1986, 'Chaos' by James P. Crutchfield, J. Doyne Farmer, Norman H. Packard and Robert S. Shaw, pp.38–49.
Discover, November 1992, 'Does chaos rule the Cosmos' by Ian Stewart, pp.56–63.
Discover, March 1993, 'Chaos hits Wall Street' by David Berreby, pp.76–84.
Wired, July 1994, 'Cracking Wall Street' by Kevin Kelly, pp.92–95, 132–136.

DAMASCUS STEEL

THE CITY OF DAMASCUS in Syria is famous for many things. Damascus is probably the oldest continuously inhabited city in the world. It was so beautiful that legend has it Muhammad refused to visit Damascus because he felt that it was not right that a person should experience paradise before he died. It was on the road to Damascus that the conversion of Saul of Tarsus to Saint Paul the Apostle began, and it was in Damascus that Cain, buried his murdered brother Abel. Damascus was the home of 'damask', a shining, lustrous patterned fabric, and Damascus was where the Christian Crusaders first came across a superior weapons metal called 'Damascus Steel'.

Damascus blades were the stuff of legend, and only the Muslim blacksmiths could make them. The steel was so hard it would hold a sharp cutting edge indefinitely, but it was flexible enough *not* to break under a blow from another sword.

In his fictional book about the Crusades, *The Talisman*, Sir Walter Scott describes a meeting in 1192 between the English king, Richard the Lionhearted, and the Saracen king, Saladin. (Saladin, the Sultan of Egypt and Syria, had captured Jerusalem in October 1187. Richard I of England, *Coeur de Lion*, commanded the Third Crusade, which had the goal of re-capturing Jerusalem.) King Richard, keen to show off his weapon's technology, cut a heavy steel mace in half with just a single blow of his huge sword. But Saladin gently drew his smaller sword across a silk cushion 'with so little apparent

Damascus Steel Did Not Come from Damascus!

The first written record of Damascus blades comes from 540 AD. Damascus steel itself came from India, where it was called 'wootz'. The little soap-sized lumps were widely traded. Damascus steels were also known to the medieval Russians, who called it 'bulat'. Damascus steel got its name both from the location where the Crusaders first encountered it (Damascus), and from the lace-like pattern (damask) visible on the blade.

Damascus – Much-Invaded City

In 1988, Damascus, the largest city and capital of Syria, had a population of around 1.3 million. Damascus is on the site of an oasis between the deserts and the mountains, on the Barada River. Because of its precious water, the Barada River was known to the Greeks and the Romans as the River of Gold, Chrysorrhoas.

Damascene cuisine is famous throughout the Arab countries, and so Damascus is often called Al-Matbakh, the Kitchen. Damascus was a major trade centre in 3000 BC, linking the Nile and Euphrates rivers and has been continuously inhabited since then at least, mostly under the rule of foreign powers. In its time, it has been ruled by Egyptians, Hittites, Aramaeans, Israelites, Assyrians, Babylonians, Persians, Greeks, Alexander the Great, Armenians, Nabataeans, Seleucids, Byzantines, Romans, and even Australians! Although it was one of the earliest cities to be converted to Christianity early in the first century AD, Damascus became Islamic in 636 AD, after it was invaded by Arab armies under the command of General Khalid ibn al-Walid. The Saracens held Damascus in the 12th and 13th centuries, while the Ottoman Turks controlled it from 1516 to 1918. Lawrence of Arabia was involved, with Arabian soldiers, in the British invasion of 1918. Another foreign power (France) controlled it under a League of Nations decree until 1941, when Syria finally became independent, the last of the French finally withdrawing in 1946.

effort that the cushion seemed rather to fall asunder, than to be divided by violence'. Scott goes on to describe Saladin's blade as 'a dull blue colour, marked with ten millions of meandering lines'.

Since they first came across it about a thousand years ago, European blacksmiths have tried unsuccessfully to make Damascus steel. Even the Muslims lost the secret of how to make a Damascus blade some two centuries ago. But just recently Oleg B. Sherby and Jeffrey Wadsworth from Stanford University think they have rediscovered the lost secret.

They reckon that it was a two-stage process, each with a few steps. In the first stage, the early Muslim blacksmiths made a steel that had about four times as much carbon as modern steels – so it was very hard, but unfortunately, brittle. In the second stage, they would reheat and 'massage' the steel to make it flexible.

As the first step in the first stage, those ancient Muslim blacksmiths made a carbon-rich steel. To do this, they heated iron ore (a compound of iron and oxygen) to a white heat (around 1200°C) with charcoal (which is basically carbon). The carbon reacted with the oxygen, removing it from the iron, and some of the carbon infiltrated into the iron. When the iron cooled down, it had a spongy texture. The iron had *some* carbon in it, but *not enough* carbon to make Damascus steel.

In the next step, the blacksmith hammered this spongy iron. This hammering removed impurities such as slag, and left behind little chunks of wrought iron the size of jelly beans. These pieces of wrought iron were put into a small clay pot (about 75 mm across and 150 mm high, roughly the size of a tin of baked beans) with some more charcoal to increase the carbon content, and the pot was sealed to keep out oxygen.

The clay pot was heated until the blacksmith heard a sloshing sound, which meant that some liquid metal was present. (At this stage, some of the carbon had diffused into the surface of the chunks of wrought iron.) The metal was then allowed to cool down very slowly. (This very slow cooling allowed enough time for the carbon to thoroughly 'soak' into the bulk of the iron lumps.) When the pot was broken, a piece of carbon-rich steel, the size of a cake of soap, was left behind.

What happened as the liquid metal cooled down below $1000°C$ was that some of the carbon came out of solution to make a network of whitish iron carbide (Fe_3C). The iron carbide made the steel hard. Under a microscope, the steel looks like chicken wire – a continuous mesh of iron carbide around darker crystals of iron. (It's this mesh of iron carbide that looks like lace-work, and gives the characteristic damask markings to Damascus steel.) Because the mesh was continuous, cracks could travel along it, making the steel quite brittle.

The first stage of turning iron into Damascus steel was finished. While most modern steels are under 0.5 per cent carbon, this medieval steel was about 1.5–2 per cent carbon. This high carbon level meant that it was very hard, and could take a very sharp edge – but now there was the problem that it was also very brittle, and would easily crack in the heat of battle.

It was now time for the second stage in the blacksmith's art. The aim was to break this *continuous* mesh of iron carbide (Fe_3C) into a collection of *separate* blobs. To do this they heated this cake of hard, but brittle, steel to a colour between blood red (650°C) and cherry red (850°C) – around 750°C. Higher temperatures would have dissolved the desired iron carbide back into the steel. They then hammered the steel vigorously, making the steel up to eight times thinner, and at the same time, breaking the **mesh** of iron carbide into

Carbon and Iron

Iron is the fourth most common element in the Earth's crust, making up about 5 per cent of the weight of igneous rocks. 'Pure irons', such as wrought iron and ingot iron, have only traces of carbon. 'High-carbon steels' have up to 1 per cent carbon, while 'ultrahigh carbon steels' have up to 2 per cent carbon. 'Cast irons' contain between 2.4 – 4.5 per cent carbon.

Famous Knife Owners

When King Tutankhamen was buried some 3400 years ago, two daggers were entombed with him. Each had a handle decorated with lapis lazuli, gold beads and turquoise. But while one blade was made of gold, the other was of iron. In the time of King Tutankhamen, the iron-bladed knife was the more valuable of ➤

▷ the two, because iron was very rare then!

A master knifesmith, Buster Warenski of Richfield, Utah, made a copy of the gold knife for a knife collector, Phil Lobred. It took five years to complete, and used 32 ounces of gold. Lobred has already refused an offer of $US 100 000 for the gold replica knife.

Sylvester Stallone commissioned a knifesmith, Gil Hibben of Louisville, Kentucky, to fashion a modern version of the classic Bowie knife for his movie 'Rambo III'. Eighteen knives were made, of which Stallone bought six. One of these knives recently sold for $US 5000!

separate **blobs** of iron carbide.

So now the steel was very *hard* (because of the iron carbide) and very *flexible* and *non-brittle* (because the iron carbide was in individual little blobs, and so any cracks that did form would stop dead, and not spread). This microstructure was then locked into place by rapidly cooling the steel in water.

But this was just a **blob** of Damascus steel, not a Damascus **blade**. The blacksmith would then take two pieces of this Damascus steel, place one on top of the other, reheat them, and weld them by hammering them together. The weld would have to be perfect, with no hot spots or cold spots, nor with any air bubbles between the two steel layers. Then the steel bar would again be reheated to the correct temperature, and folded in half, and the two halves welded by more hammering. Now the steel bar had four layers. Another eight such folding operations would give you a blade with 1024 layers – and one single air bubble at any stage could ruin the whole blade! But the finished blade was now very strong – after all, a slab of laminated plywood with its several layers is stronger than an identically sized slab of plain timber.

The hundreds of meandering white lines on the dull blue background were a fairly reasonable medieval indicator of quality control. The blade was first polished and then etched with an acid that attacked the iron, but left the whitish iron carbide untouched. The pattern became visible only if the blade had a very high carbon content (for hardness) and had been well forged (for flexibility).

There are many patterns. One pattern with regular bands was called 'Muhammad's Ladder', because dead soldiers would climb it to get to heaven. 'Ocean Waves' is a random pattern that looks like ripples in the water, while 'Maiden's Hair' has alternating patterns that look like rosettes, and a woman's hair.

With modern technology, we can apply more subtle tests of quality other than looking for such patterns. We now know that the pattern is visible only if the iron carbide grains are

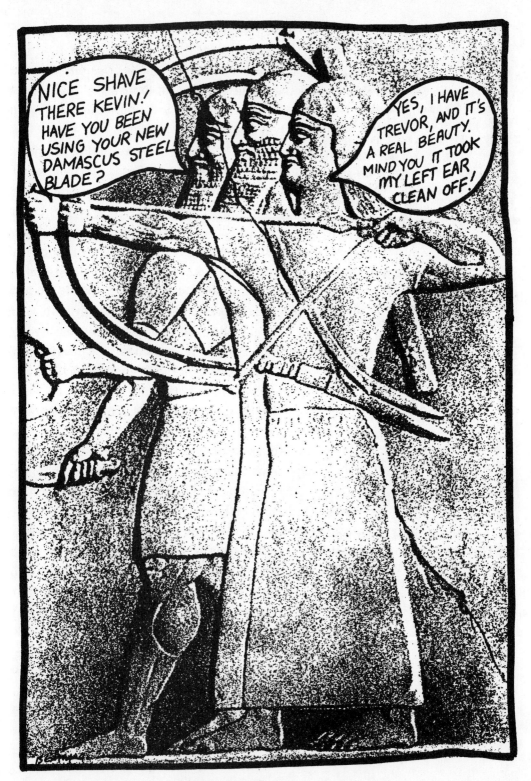

Body of a Slave...

In the Balgala Temple in Asia Minor a description was found of how to harden the bulat (Damascus steel): 'The bulat must be heated until it does not shine, just like the sun rising in the desert, after which it must be cooled down to the colour of the king's purple, then dropped into the body of a muscular slave...the strength of the slave was transferred to the blade and is the one that gives the metal its strength.'

This description may be romantic, but it does not give the best Damascus steel. The colour of 'sun rising in the desert' corresponds to a temperature of around 1000°C – much too high, as it would allow the iron carbide to dissolve back into the steel. However, on being air-cooled to 'the colour of the king's purple', around 800°C, the iron carbide would re-emerge. Unfortunately, the iron carbide would be in a coarse network, not a fine-grained one. Quenching it in 'the body of a muscular slave' (around 37°C) would have locked in this coarse network.

The sword or knife made to this recipe would have been hard, but brittle. And of course, stabbing a slave would have been cruel and wasteful.

relatively large and coarse. If the iron carbide grains are very small, which means that the blade will be harder and more flexible, the pattern is not visible to the naked eye. Even so, those ancient blacksmiths did very well with the technology they had available.

Damascus steel is now attracting a lot of attention – NASA Ames Research Laboratory, Lawrence Livermore National Laboratory in California, Caterpillar and North Star Steel in Minneapolis are all working on it. In their research, they've made some varieties of Damascus steel that as well as being resistant to corrosion are also 'superplastic'. While modern steels will crack if they are stretched by more than 50 per cent, these superplastic steels can (when heated to 600°C) be stretched like plasticine to over 10 times their original length. They can also be squeezed like putty into moulds, to make complicated parts like gears. This saves money, because getting the metal close to the final desired shape means that you don't need so much expensive machining. The Caterpillar company, which makes bulldozers, is already testing Damascus steels – probably for super-hard bulldozer blades, and teeth that will not shatter.

The knifesmiths are also working with a modern version of Damascus steel, and their products are now hitting the market. The head of the Russian secret service now carries a Damascus fighting dagger made by Bud Nealy. These modern Damascus blades are amazing. A blacksmith can sharpen a blunt Damascus blade to a shaving edge with nothing more than the back of a leather-bound book! Because the steel is so hard, it will hold this edge even after sawing through a 15 centimetre tree trunk. And one

Damascus – Religious City

Both the Koran and the Bible cite Damascus as being connected with the first murder in recorded history! Cain killed his brother Abel, and after carrying him for several days, finally buried him on a rocky hilltop, some kilometres out of Damascus. An-Nabi Habil, the tomb of Abel, is now sacred to the Druzes.

Saul of Tarsus, who had plagued the Christians in Jerusalem, was travelling to Damascus to continue his persecution of them. He was near the village of Darayya, when he was blinded by a vision of the Lord, and was taken from there into the city of Damascus, to the house of a Christian. Acts 9:10 & 11 say: 'And there was a certain disciple at Damascus, named Ananias: and to him said the Lord, in a vision, Ananias. And he said, Behold, I am here, Lord. And the Lord said unto him, Arise, and go into the street which is called Straight...'

The 'street called Straight' still exists today, but it is quite different from the 30 metre wide, mile-long street which the Romans called Doconomos. It's now a narrow, twisty street that winds its way between the shops that encroach upon it from each side, and the original street level is some 5 metres under its current level.

Saul made so many enemies in Damascus, that he had to be smuggled out over the city walls, lowered in a basket!

A Christian cathedral, the Cathedral of St John the Baptist, was built in Damascus, and is said to hold St John's severed head. In 636 AD, Damascus came under Muslim rule, and the cathedral was divided into two quite separate sections – one for the Muslims, and the other for the Christians. In 705, the Umayyads turned it into a wholly Muslim mosque. It is now the fourth most holy mosque in the Arab world – and still reputedly has the head of St John the Baptist. There has been some kind of religious building ('pagan' temple, church, mosque etc) on that site since around 4000 BC! Even the Romans had built a Temple to Jupiter on the site.

The Umayyads were remarkable rulers, and created an Islamic empire that was larger than the Roman Empire! It stretched from the Pyrenees to the Indus River in India.

The Tomb of Al-Sitt Zainab, the granddaughter of Muhammad, is some 5 kilometres to the south of Damascus.

witness describes how a single swipe with a 20 centimetre Damascus blade cut cleanly through four free-hanging Manila hemp ropes, each 25 millimetres thick! As well, the blade is so flexible that it can be bent to incredible angles without breaking.

With so many uses springing up for this ancient flexible metal, perhaps the wheels of industry might have to steel themselves for a back-to-the-future revolution.

References

National Geographic, April 1974, 'Damascus, Syria's uneasy Eden' by Robert Azzi, pp.512–535.
Scientific American, February 1985, 'Damascus Steels' by Oleg D. Sherby and Jeffrey Wadsworth, pp.94–99.
Van Nostrand's Scientific Encyclopaedia, 1989, Van Nostrand Reinhold, pp.1599–1617.
Popular Science, April 1993, 'Damascus steel lost and found' by Sandy Fritz, p.35.
Fortune, 16 May 1994, 'Back to the future with flexible steel' by Alison L. Sprout.

FUNGUS IS FAMILY

ON OUR PLANET TODAY, life has evolved from simpler creatures into creatures as diverse as wheat and whales, mushrooms and viruses. Biologists classify today's life into various kingdoms such as plants, animals, fungi and so forth. They have worked out roughly when each major kingdom split off from various common ancestors. But everybody was surprised when recent research showed that we humans are more closely related to a packet of baker's yeast than we are to the plants!

Baker's yeast and brewer's yeast are just two examples of the incredibly diverse world of fungi. There are about 100 000 different species of fungi, and these have adapted incredibly well to all niches of life. One fungus, the so-called 'kerosene fungus', *Amorphotheca resinae*, can live quite comfortably in the tanks of jet fuel (which is basically kerosene). So long as there is a small amount of water in the tank, it can eat the kerosene!

A fungus is in fact the world's largest living organism. One fungus, *Armillaria ostayae*, discovered in 1992 in the state of Washington in the USA, covers an area of around 6 square kilometres! The *largest* edible fungus is a puff ball, *Langermannia gigantea*, that was discovered in Yellow Springs, Ohio – it's about 63 centimetres across. But that was the *largest* edible fungus, not the heaviest. The *heaviest* edible fungus was a 'chicken of the woods' mushroom, *Laetiporus sulphureus*, found in the New Forest in Hants on 15 October 1990, weighing over 45 kilograms!

The most poisonous fungus is the yellowish-olive death cap, *Amanita phalliodes*, which causes 90 per cent of all deaths from fungi. A lethal dose is around 50 grams of the fresh mushroom. This fungus killed Cardinal Giulio di' Medici, and Pope Clement VII. Pope Clement VII was so fond of mushrooms that before he died on 25 September 1534, he actually passed a law forbidding others from eating any mushrooms found growing in the Papal States.

The fastest-growing creature on our planet is the stinkhorn fungus, *Dictyophora*, which lives in the tropical forests of Brazil. It can grow at 5 millimetres per minute, and reach full size in just 20 minutes! When it grows, it absorbs water so rapidly that with your naked ear, you can hear crackling noises as the water rips apart tissues inside this fungus!

One special thing about fungi is that they get their food by directly absorbing the nutrients – they don't have a mouth. They absorb their food through a set of fine and delicate filaments called 'hyphae'. When these hyphae have spread out into a big mass, the mass is called the 'mycelium'. The hyphae release enzymes to dissolve any food they find, which is then absorbed directly through the walls of the hyphae. When you're eating a mushroom, you're actually eating the mycelium of a fungus that has developed into the large reproductive 'fruiting body'. It's this fruiting body that leads to the next generation of fungi.

Mushrooms are among the more popular fungi, and are actually much bigger than they look. The bit that you pick and eat is only about 10 per cent of the weight of the fungus. The rest of it lives underground as a network of tiny strands. These strands feed the fruiting body, which usually lives above the ground. Some mushrooms will soak up nutrients from the roots of trees and, in return, help the tree soak up nutrients as well.

But if mushrooms taste good, then another fungus, the truffle, tastes like heaven. This small warty fungus is called the French Périgord truffle. Tiny slices of this French truffle are mixed into pâté de foie gras. It's very precious. Black

Shaking Ants Leave it to Fungus!

Have you ever tried to cut a sheet of floppy paper with scissors that weren't quite tight enough? The paper just slips between the blades. But if you then use the same scissors to cut some stiff paper, the blades will slice easily through the paper. One type of ant uses this as a trick to cut leaves – it makes the leaves stiffer, and easier to cut, by vibrating them!

The ant is the parasol ant. As each ant marches along, it holds up a leaf, which looks like a miniature parasol or umbrella. But the leaf is not there to protect the ant from the Sun – it's for food. The parasol ants take the leaves back to the nest, chew them into a soft mush or pulp, and then activate the mush by adding a special liquid that is loaded with nutrients and fertilisers. They add the activated mush to a much bigger pile of mush, where a specially cultivated fungus is already growing. Finally, they eat the fungus.

This is a beautiful two-way relationship. The fungus needs the ants, because it does not survive very well at all in the wild. The ant needs the fungus, because it can't eat anything else.

According to some scientists, the parasol ants have been cultivating the very same fungus for over 20 million years. Even today, these ants take their fungus cultivating very seriously indeed. Scientists looked at one nest which had been occupied by ants for some six years. The gardens in which the ants grew their fungus were over 4 metres underground. There were over 40 000 ants living in that single nest, and during that six years, they had dug out nearly 2000 chambers for cultivating the fungus, and had shifted some *40 tonnes* of dirt! And in the above-ground area around the ant nest, the scientists found that in one season alone, the ants had cut about 6 tonnes of leaves off the trees, to feed their pet fungi. These fungi must be delicious, because the Ambrose beetle, and various species of termites, also cultivate them for food.

To cut a leaf, the ant first hooks herself onto it with her hind legs. At the other end of her body, on her head, she has mandibles, which are like giant cutting blades. The edge of each blade is serrated, and is hardened by a coating of zinc! She starts by placing one mandible against the edge of the leaf, and the other mandible on the surface of the leaf. She then begins to rub a file on one part of her body against a scraper on another part. These vibrations travel through her whole body – *and* through her mandibles. The mandibles are vibrated at around 1000 vibrations per second – in all three directions at the same time, *not* just up and down. She cuts through the leaf by bringing the mandibles together, and will have to make about 35 separate bites to successfully cut out a single section for feeding to the hungry fungi.

Now there are two separate effects going on in this cutting process. The less important effect is that the vibrated mandible behaves like an ➤

▷ electric knife with a serrated edge. But the more important effect is that the vibrations stiffen the leaf, making it easier to cut. This is very handy when the ants have to cut a nice young juicy floppy tender leaf. In fact, the ants will do this vibrating thing 70 per cent of the time when cutting through a tender leaf, but only 40 per cent of the time when cutting through an older, tougher leaf.

Now human flesh is very floppy, and this same stiffening effect is used in modern medicine to cut very thin slices of human tissue pathologists can look through with a microscope. They use a machine called a 'vibratome', which has its blades following an elliptical (or egg-shaped) path. But the parasol ants have known this trick for millions of years. By shaking their thing, the ants keep a good supply of nutritious fast food always on hand!

Périgord truffles cost around $1000 per kilogram, while the rarer white Périgord truffles cost up to $3000 a kilogram.

Truffles are so important that there is even a Truffle Fraternity. Every January, in their ceremonial robes, members attend the Truffle Mass, which is held at the 12th century church in Richerenche. Instead of putting money in the collection basket, they donate truffles! In 1992, the collection basket at the end of the service in this church held some 4.5 kilograms of truffles worth around $3000.

This fungus grows underground near the roots of oak, hazel and beech trees. It starts off almost white when young, and gradually darkens with an irregular marbling as it gets older. It's quite unusual in that its fruiting body lives beneath the surface of the soil, and this makes it very difficult to find. Usually it can be found only by the smell that it gives off. It is this exquisite odour, which has been called 'the mesmerising odour of a pig in rut', along with its musky flavour, that have made it very popular with gourmets since the times of the Romans.

There are few other clues for the dedicated truffle-hunter. If very large, it might crack the soil. Sometimes the truffle hunters are lucky enough to see small yellow truffle flies. These flies lay their eggs on the truffle, and in turn, carry the truffle spores to new areas. But its smell remains the best way to track the truffle down.

Even though the truffle takes seven years to come to maturity, it's at its peak for just one week – so there is some urgency involved in finding and picking it! Traditionally, pigs and dogs have been used to track down the truffles by following their distinctive smell. Many truffle hunters are switching over to dogs, because the pigs are so heavy that they will often crush the underground truffle accidentally. Furthermore, pigs are rather sly, and will try to eat the truffle, if not closely supervised! The dogs aren't really interested in eating truffles.

It is because of their scarcity that truffles command such a high price. Human ingenuity has taken three different approaches to this problem – domesticate the truffles, created a 'synthetic' truffle sauce, or look for a better way to find truffles.

Domesticating the truffle, so that it grows happily in a nursery, is harder than it seems. There are occasional reports from the USA and Japan that yet another team of biologists has succeeded, but the domesticated truffles have not yet made it to market.

Following a different line of research, a team of French biologists claims that it has identified the chemicals that give the truffle its special aroma. If they can make this group of chemicals, they will market it as a cheap synthetic truffle sauce.

The third approach has been taken by scientists from the University of Manchester Institute of Science and Technology, who have cooperated with French scientists from Toulouse to invent and market a portable truffle detector. This detector has 24 electronic sensors. Inside each sensor is a different polymer (a big chemical made up of many smaller chemical sub-units that are often identical), which is sensitive to different smells. When the smell chemicals float through the air from the truffle, and land on these polymers, the electrical resistance of these polymers changes. Different odours will stimulate different combinations of these 24 separate sensors. These sensors are so sensitive that they can pick up

Our planet is about 4.6 billion years old. Our earliest fossils date back to 3.8 billion years old. That creature was a single-celled creature, a stromatolite. Life existed as single-celled creatures until about 1 billion (1000 million years) ago, when the first multi-celled creatures began to appear. The first creatures with hard bodies (like shells) appeared around 600 million years ago.

Life came on to the land about 400 million years ago and reptiles and dinosaurs were roaming the earth around 200 million years ago. The dinosaurs died out about 65 million years ago and by 55 million years ago mammals had expanded into most ecological niches.

The first primate (two-legged mammal) appeared about 50 million years ago. Three million years ago, our ancestors were a bit smaller than we are today, but had a brain about half the size of ours.

individual chemicals when they are diluted to concentrations as low as several parts per billion! One particular combination of electrical resistance of these 24 different sensors signals that a truffle is nearby, and this is the signature that the scientists have programmed into a microprocessor. At the moment, the pigs and dogs harvest only 80 per cent of the possible crop. With a truffle detector they should be able to get the full 100 per cent.

Besides tasting good, fungi do a lot for the benefit of us humans. Fungi make our bread and beer for us, via baker's yeast and brewer's yeast. Soy sauce comes from soya beans, but it has to be fermented. Two fungi are used to do this job, *Aspergillus oryzae* or *Aspergillus soyae*. And *Aspergillus oryzae* also makes the chemical called amylase, which is used in fermenting alcohol. Liquid glue is made with a family of chemicals called proteases, which comes from a fungus called *Aspergillus flavus*.

A Norwegian fungus, *Tolypocladium inflatum*, gives us the class of drugs called cyclosporines. These drugs are very useful to people who have had a transplant of an organ (such as a kidney, heart or liver) from another person. Ordinarily, their immune system would reject this foreign organ, but the cyclosporines suppress their immune system, and allow the transplanted organ to survive.

A fungus called *Cephalosporium acremonium* gave us an entire class of antibiotic drugs called cephalosporins. In 1948, Giuseppe Brotzu, a professor of bacteriology in Sardinia, discovered this fungus living near the mouth of a sewage outfall from the city of Cagliari!

One of our very first antibiotics, penicillin, comes from a fungus called *Penicillium notatum*. Other useful fungi in the penicillium family are *Penicillium roqueforti*, which makes Roquefort cheese, and *Penicillium camemberti*, which ripens Camembert cheese. You can even get vitamin B from fungi – simply irradiate ergosterol, which comes from waste brewers' yeast.

One particular fungus is being used to kill insects. Locust

Sheep Stomach Fungus

Fungi from the stomachs of various exotic animals look like being able to help our Australian grazing animals. For a long time biologists have known that various African animals were very efficient at digesting grasses, especially when they were dry and unappetising. So a team led by Dr Geoff Gordon at the CSIRO Division of Animal Production began looking inside the stomachs of various exotic animals for any clues.

A sheep has two stomachs to help it digest its food. They found that a single millilitre of liquid from the first stomach of the sheep contained a lot of living creatures — about a billion bacteria, a million protozoa and over 10 000 fungi.

Fungi have a natural ability to break down plant carbohydrates and cellulose. They can actually digest the fibre and cellulose in dry grasses, and break it down into other carbohydrates that are easier for the sheep to digest. So these scientists collected fungi from the stomachs of grass-eaters all over the planet — animals like the water buffalo, domestic cattle, kangaroo, deer, elephant, giraffe, oryx and even eland. They found many different types of fungi living in the stomachs of these animals, but they decided to concentrate their research on just five main types.

Now when times are bad, the feed is dry, shrivelled-up grasses. These take a long time to digest. According to Dr Gordon, the grass 'forms a bottleneck in the sheep's stomach. This usually prevents the sheep from eating enough to sustain themselves.'

In one experiment, they actually removed the usual fungi from sheep, and found that the amount of food that they ate went down by 30 per cent. When they gave the fungi back to the sheep, their food intake went back to normal. And when they added some of the fungi from the exotic grass-eaters, they found that the food intake went up by 10 per cent. In fact, they think that if they were to fine tune the fungi a little bit, they could even increase the food intake by 20 per cent.

Now this technique would be very useful if you knew a drought was coming. You could simply add some of these exotic fungi to your sheep's diet just before the food shortage was about to occur, and the sheep would be able to get more nutrition from the reduced amount of food they did have access to. This would reduce the amount of grazing that the sheep had to do, and ultimately, would reduce the degradation of the land. Looking at it another way, the farmers could get more meat and wool from fewer animals.

plagues are a big problem in Africa, and between 1985 and 1989 alone some $400 million was spent trying to eradicate locusts. Most of this money was spent on the organophosphorous insecticide Fenitrothion. Over 11 million litres of this chemical was sprayed! There have been strong environmental concerns about spraying such huge quantities of an organophosphate over large areas.

So Chris Prior at the International Institute of Biological Control, which is in Berkshire in the United Kingdom, came up with the idea of spraying the locusts with a living fungus! The fungus they have been using is *Metarhizium flavovride*. This fungus doesn't bother humans, and in fact, will infect only insects.

Once a spore of this fungus has landed on an insect, it will sprout, sending tubes (or hyphae) into the hard shell (or cuticle) of the insect. These tubes are about 20 microns long (about one-third the thickness of a human hair) but very skinny, only one or two microns thick. Once each tube has penetrated into the cuticle, its end swells up, and it continues to force its way deeper into the cuticle. It's helped by enzymes which leak out from the end of the growing tip of the hyphae of the fungus. Once it breaks through the cuticle, it grows like crazy and just multiplies until it has clogged up the entire insides of the insect! The fungus kills the insect simply by local pressure effects, not by poisoning it with any fancy toxins.

This fungus can be sprayed through the same spraying equipment used for the organophosphorous insecticide. The fungus is mixed into an oily liquid, and is present as a suspension of millions of tiny spores. These tiny spores can pass easily through the nozzles of the spraying equipment.

Locusts have been living with this fungus for millions of years, but they are only very occasionally infected by it – usually only during the wet season. Those parts of Africa where the locust plagues occur are dry for practically all of the year, and there is only a brief wet period during which the insects are infected. So spraying the fungus as a liquid into the insects is one good way to infect them. There's one great benefit to using a living creature (such as a fungus) to control another living creature we don't want, such as a locust. The insects breed rapidly, and so they can quickly develop resistance to any nasty chemicals we want to spray on them. The advantage of infecting them with a living creature is that as the locusts try to breed a resistance to the fungi, the fungi will evolve right along with the locusts, and try to overcome this resistance.

Now it is still early days, and we don't know how effective this fungal spray will be, even though trials in Benin, Mauritania and Niger have been most successful. But if it does work, we can then start looking for fungi that infect pests such as caterpillars, thrips and various sap-sucking bugs.

A fungus from the stomach of a sheep can not only stretch the life of city landfills, but it also give us 'stone-wash' jeans. This research was done by a team led by Dr Jim Aylward from the Australian Commonwealth Scientific and Industrial Research Organisation (CSIRO) Division of Tropical Crops and Pastures. The team isolated a fungus from the sheep stomach, and cloned genes from that fungus. These genes made extremely powerful and useful enzymes called cellulases, which can dissolve carbohydrates. They found that these cellulases from the stomach of a sheep were the most powerful chemicals for dissolving fibre ever discovered! In a diluted form, these cellulases could make 'stone-washed' jeans.

Their fibre-dissolving property also meant that they could help with the problem of the city landfills. Forty per cent of the domestic garbage that goes into a city landfill is lignocellulose – cardboard, wood and paper. These enzymes, by breaking down this lignocellulose into carbon dioxide (the infamous greenhouse gas) and water would free up lots of room, and so stretch the life of these city landfills (although it would perhaps not be so good for our air).

So fungi may be as useful to us in the future, as they have been in the past. And as we look into the past, we are discovering that we might be closely related to the not-so-humble fungus. The research involved has been carried out by Patricia Wainwright and her colleagues from the Institute of Marine and Coastal Sciences at the Cook Campus of Rutgers University in New Brunswick in New Jersey, and from the Center for Molecular Evolution at the Marine Biological Laboratory in Woods Hole, Massachusetts. They were interested in finding out about the ancestry of various living creatures, so they analysed various genes from creatures such as algae, sponges, plants, animals and fungi.

Now genes will change and mutate with time, and there are various reasons for believing that they mutate at a fairly constant rate. The scientists compared the similarities and the differences between these various species, looking at a few specific genes. All things considered, we are probably closer (in an evolutionary sense) to a creature which has genes that are similar to our genes – and probably further away from a creature which has genes that have mutated and changed to be different from our genes. The standard theory is that all current living creatures evolved from a single-celled creature some 3 billion years ago, and that various kingdoms split off after that. It's the timing of various branchings that is still contentious.

Patricia Wainwright and her co-workers found that animals (including humans) and fungi share a common history of evolution. Our common ancestor probably had characteristics of both plants and animals. In fact, they also claim that our common branch of the evolutionary tree split off from the plants about 1.1 billion years ago.

To summarise, they say that our ancestor first split off from the plants, and then later split into animals and fungi – so we animals are close to the fungi. This might explain why drugs that are used against fungi generally have nasty side effects for us humans. It could be because we are genetically close to the fungi, and so any drug that affects fungi, will also affect us. In fact, some patients call the anti-fungal drug Amphotericin 'Ampho-terrible', because of its side effects!

Fungi Help Trees

Some fungi live in a happy marriage with the roots of certain trees. The fine filaments (the hyphae) of the fungus surround the tips of the roots of the tree. It's a two-way deal. The tree carries out photosynthesis, and with the energy thus produced, makes various sugars. The fungal hyphae absorb these sugars, and in return, the fungus concentrates phosphorus and nitrogen around its hyphae. As the tree 'drinks' water, it takes up these, and other, nutrients.

The most valuable timber tree in the Northwest of the USA, the Douglas fir, is 'married' to about 2000 different types of fungi!

Our shared ancestory with fungi is probably bad news for those with a delicate ego who would rather be related to the giant oak tree than the mildew on their shoes in the wardrobe, or the athlete's foot between their toes! On the other hand, vegetarians shouldn't worry about mushrooms being meat – after all, we did say goodbye to the fungi about a billion years ago!

References

Scientific American, September 1981, 'The microbiological production of food and drink' by Anthony H. Rose, pp.95–104.

Science, Vol 260, 16 April 1993, 'Animal-fungi link', p.273.

Science, 16 April 1993, 'Monophyletic origins of the metazo: an evolutionary link with fungi' by Patricia O. Wainright, Gregory Hinkle, Mitchell L. and Shawn K. Stickel, pp.340–342.

New Scientist, No. 1916, 12 March 1994, 'Plague of fungus could limit locusts' by Judith Perera, p.15.

New Scientist, No. 1956, 7 December 1994, 'Conservative ants grow what they know' by Kurt Kleiner, p.15.

Science, Vol 267, 6 January 1995, 'Use of a sound-based vibratome by leaf-cutting ants' by Jürgen Tautz, Flavio Roces and Bert Hölldobler, pp.84–87.

New Scientist, No. 1959, 7 January 1995, 'Electronic sniffer stalks famous fungus' by Chris Wilson, p.17.

LANGUAGE IN JUNK DNA

FROM TIME TO TIME in the news, you might have noticed reporters getting excited about newly discovered **genes** – a gene for breast cancer, a gene for a muscle disease, and so on. These genes are part of our DNA. The reason that the reporters are getting excited is because of the possibility of using this information to cure, or even prevent, various diseases.

Now DNA is the very stuff of life itself, so it's a pretty big topic (one of the *big* ones, in fact). So let's talk about DNA first, and what genes are, before we finally discuss the hidden language in the DNA.

First, let's discuss DNA – the blueprint for life. Now, 'DNA' stands for **d**eoxyribo**n**ucleic **a**cid. It's a very big chemical. It looks a lot like an ordinary ladder. There are two side rails (one on each side) and many rungs joining these side rails. It's a very long ladder – there are about 3 billion rungs. If you could tease it out, it would be a few metres long! But thanks to some very clever packing, the DNA is all coiled up very tightly to fit inside something as small as a cell. This very long ladder of DNA exists in each cell in our body (except for the red blood cells).

This next bit is complicated, so it's no wonder that it took a few Nobel Prizes to work it out. DNA is a blueprint. The biochemical 'machinery' to 'read' the blueprint of the DNA is present in

The Genetic Code

One great discovery has been made concerning the four different types of rungs, A, C, G or T, that make up the 3 billion rungs in the DNA ladder of life. A set of three rungs contains all the information needed to tell the biological 'machinery' in the cell to make one amino acid. So the set of rungs AAA will make one amino acid, while the rungs TTT will make a different amino acid, and so on. There are 64 possible combinations of rungs, but only about 20 different amino acids. So sometimes six different combinations will code for the same amino acid.

Genes and Disease ①

Genes can be involved in inherited diseases, and sometimes different people can have different versions of the same gene. So on average, one person out of every 70 people will carry a version of the gene that can pass on the disease called 'cystic fibrosis'. On average, 69 out of every 70 people won't carry this version of the gene — they'll carry a different version of this gene.

If you carry one copy of the 'cystic fibrosis' gene, you won't suffer from the disease, but you will be a carrier. If your partner does not carry this gene, none of your children will have the disease, but (on average) half of them will be carriers. If your partner does carry the gene, (on average) one-quarter of your children will have the disease.

One odd thing about cystic fibrosis is that it seems to be mainly a disease of Caucasians. It seems to be very rare (or absent) in Chinese people. If you know you are a carrier for cystic fibrosis, make your ➤

practically every cell of your body. Once this 'machinery' has read the DNA, it 'instructs' other biochemical machinery to make amino acids. Now the special thing about amino acids is that if you join amino acids together, you have a protein. In other words, DNA 'makes' proteins.

What happens next depends on the particular cell the DNA is inside. Proteins are used to make the cell membrane, as well as many structures inside the cell. If the DNA is in a **skin cell**, the DNA can deliver instructions to split that skin cell into two more skin cells – to replace other skin cells which have died. If it's in a cell in your **pancreas**, the DNA can give instructions for that cell to make the protein hormone called 'insulin'. In **muscle** cells, the proteins can end up as muscle fibres.

And if the DNA is inside a **fertilised human egg**, it will give instructions that use the next nine months to make a baby. In this case, the egg has half the DNA needed to make a full human being, while the other half of the needed DNA is in the sperm. The fertilised egg has all the DNA required.

And a **gene**? Well, that's just a large number of these rungs that do a *specific* job, like making nerves, or making collagen. The average gene is about 600 rungs long and can 'code' for about 200 amino acids. In the mid-1990s it seems that there are about 100 000 different genes in the human DNA. In fact, the genes make only about 3 per cent of the known DNA. The remaining 97 per cent (the 'junk DNA') does not code for amino acids, and does not seem to have any known function. In other words, genes are the 'useful' 3 per cent, while the remaining 97 per cent has (at least in the mid-1990s) no known function.

The scientists exploring the DNA have always been a little uneasy calling 97 per cent of the known DNA 'junk'! Luckily, other scientists have come to the rescue, and have found strange hints of a language hidden in this so-called 'junk DNA'.

We know that there are about 3 billion rungs on this ladder of life, the DNA. But it's only in the last few years that we have begun to try to map them all, through the Human Genome Project. The word 'genome' just means the sum total of all of the rungs of the ladder. The Human Genome Project will be finished around the year 2010, but already we have discovered a few interesting genes, like one of the genes for breast cancer, and a few of the genes for various obscure and rare diseases.

There is a general rule which applies in many fields of science – 'The Map defines the Territory'. So the real excitement will come once we have mapped the entire human DNA.

Let me explain how the Map defines the Territory. Around the turn of this century, we began to have the first accurate maps of the continents and the off-shore underwater continental shelf. In the early 20th century, a geologist, Alfred Lothar Wegener, made little paper copies of the continents, and found that, for example, the two coastlines of the Atlantic Ocean would slot into each other. He thought that perhaps they had been joined together some time in the past. He began to think about the possibility that perhaps long ago the continents had been in different locations. He gave a lecture on this topic of 'continental drift' in 1912. And so today we have the theory that the continents drift around the surface of our planet, sometimes running into each other and sometimes moving apart. Because Wegener could look at the whole map of the Earth, he could get a different viewpoint, and this led to a new theory.

Another example of the Map defining the Territory is the Periodic Table of the Elements, devised by the Russian scientist Dmitry Ivanovich Mendeleyev. In 1869, when not all of the natural elements had yet been discovered, he began plotting the elements on paper, according to their increasing atomic weight. He noticed that he could arrange them in columns and rows, so that the elements in each row all had similar properties. He also noticed that there were gaps in this table. So he took a guess and said that each gap was the 'home' of an undiscovered element. He also said that the

▷ partner a Chinese person, and you can just about guarantee that your children won't have the disease!

DNA is Big!

There are 3 billion rungs in the DNA ladder of life. If they were written into a book, it would take one-third of your life just to read them all! In fact, a new branch of science, informatics (a collaboration between computer science and molecular genetics) was invented to be able to read and use this enormous amount of information. If we don't make the data accessible, this vast potential treasure trove could be just a big binary junkyard!

Genes and Disease ②

From time to time, a gene that is involved in a particular disease is discovered. Often, the print/radio and TV media will claim that *the* gene for this particular disease has been discovered. They're partly right, and partly wrong.

So far, very little of the DNA has been mapped (only about 5 per cent in 1995). Many diseases can be caused by more than one gene. There could be other genes for the same disease in the remaining 95 per cent of the DNA. The reporters have wrongly interpreted the scientists' work.

One case in 1995 involved the discovery of a gene for breast cancer, which was reported as *the* (only) gene for breast cancer. A few weeks later, another gene for breast cancer was found! This time, it was reported as the discovery of *another* gene for breast cancer.

properties of this unknown element should be the average of the properties of the elements above and below it, and to the left and the right of it. In 1871, he predicted the existence, and properties, of several elements that had not yet been discovered. In 1875, Paul Emile Lecoq de Boisbaudran discovered gallium; in 1879, Lars Fredrik Nilson discovered scandium; and in 1886, Clemens Winkler discovered the element germanium. Mendeleyev had been proved correct! This is another case of the Map defining the Territory.

In the same way, once we can look at the whole map of the blueprint of life, the DNA, we will then be able to move on to a deeper level of understanding and make new discoveries. At the moment, we can only guess what discoveries we might make.

Intron is one of the words the molecular biologists use to name 'junk DNA'. Introns are like enormous commercial breaks, or advertisements, that interrupt the real program – except that they take up 97 per cent of the broadcast time! Introns got Richard Roberts and Phillip Sharp (who did much of the early work on introns back in 1977) a Nobel Prize for their work in 1993. But even today, in the mid-1990s, we still don't know what introns are really for.

Simon Shepherd, who lectures in cryptography and computer security at the University of Bradford in the United Kingdom, took an approach that was based on his line of work. He looked upon the 'junk DNA' as just another secret code to be broken. He analysed the way in which the rungs of the DNA are arranged in the introns. He claimed that this arrangement is not completely random – it is actually slightly predictable. From his background in data analysis, he came up with a new theory. He now reckons that *one* probable function of introns is to act as some sort of error correction code – to fix up the occasional mistakes that happen as the DNA replicates itself.

After all, as he said to Andy Coghlan from the *New Scientist*, 'the cell puts a huge amount of its energy into the creation of these introns, then discards them...nature would

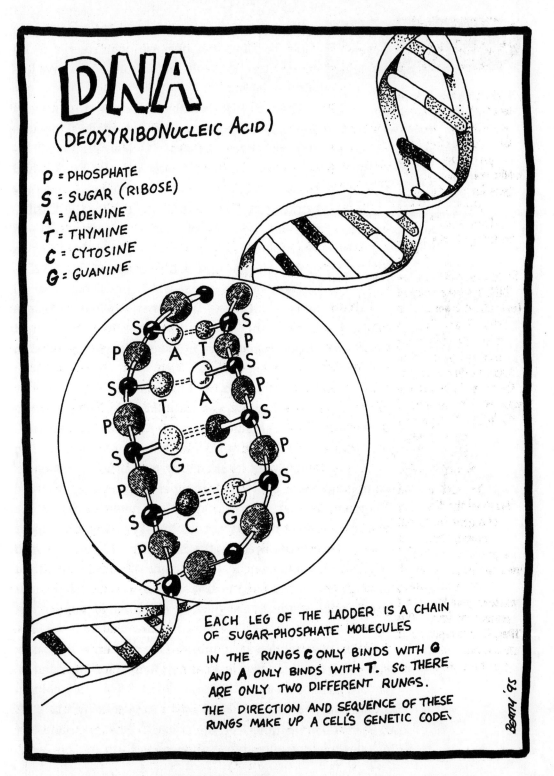

DNA

(DEOXYRIBONUCLEIC ACID)

P = PHOSPHATE
S = SUGAR (RIBOSE)
A = ADENINE
T = THYMINE
C = CYTOSINE
G = GUANINE

EACH LEG OF THE LADDER IS A CHAIN
OF SUGAR-PHOSPHATE MOLECULES

IN THE RUNGS C ONLY BINDS WITH G
AND A ONLY BINDS WITH T. SO THERE
ARE ONLY TWO DIFFERENT RUNGS.

THE DIRECTION AND SEQUENCE OF THESE
RUNGS MAKE UP A CELL'S GENETIC CODE.

BEATTY '95

Why do We Get Wrinkles?

If the DNA is still in each of our skin cells, why do we get wrinkles on our face as we get older?

While you are still growing, the blueprint inside each of the skin cells on your face tells those cells to make the chemicals elastin and collagen. These chemicals plump up your skin, and give your face that nice unlined look. (Some people will pay to have collagen injected into their face. It's a cheap, but temporary, 'face lift'.)

So why do you get wrinkles? After all, the DNA, the blueprint of life, is still there – and in good condition. But something (and in the mid-1990s, we don't know what) tells the machinery in your skin cells that reads the blueprint to stop working. So sometime in your 20s, even though the 'instructions' to make collagen and elastin are still present in each skin cell, for some unknown reason the machinery switches off, and the wrinkles of wisdom and time begin to appear.

not go to all that trouble without reason'. He could be right that one use of introns is as an error correction code, but introns could have lots of other uses.

The next big breakthrough came from a really unusual collaboration between medical doctors and physicists. Rosario N. Mantegna and his colleagues from the Center Polymer Studies and Department of Physics at Boston University worked with A. L. Goldberger from the Cardiovascular Division of Harvard Medical School. They found even more evidence that there was an 'order' of some sort in the so-called 'junk DNA'.

They applied a test which finds 'order' in all human languages. When they applied this test to introns, they found a similar 'order'! This test suggests that there is a secret language buried in the introns, the 'junk DNA'.

According to the linguists, all languages obey Zipf's Law. It was formulated by the American statistical linguist, George Kingsley Zipf, in 1947. It's a really weird law, but it's not that hard to understand. Start off by getting a big fat book. Then count the number of times each word appears in that book. You might find that the number one most popular word is 'the' (which might appear 30 000 times), followed by the second most popular word 'of' (which might appear 15 000 times), and so on. Right down at the bottom of the list, you have the least popular word, which might be 'elephant', which might appear just once.

Set up two columns of numbers. One column is the *order of popularity of the words*, running from '1' for 'the', and '2' for 'of', right down to '1000' for 'elephant'. The other column counts *how many times each word appeared*, starting off with 30 000 appearances of 'the', then 15 000 appearances of 'of', down to one appearance of 'elephant'.

Word	Order	Number Of Appearances
the	1	30 000
of	2	15 000
to	3	10 000
...	...	...
which	50	600
get	60	500
...	...	....
elephant	1000	1

If you then plot on logarithmic graph paper the *order of popularity* of the words against the *number of times each word appears*, you get a straight line! Even more amazingly, this straight line appears for every human language – whether it's English or Egyptian, Eskimo or Latin!

Now, the DNA is just one continuous ladder of squillions of rungs. These rungs are not all identical. There are four different types, called 'A', 'C', 'T' and 'G'. Since the ladder is not neatly broken up into individual words (like a book), the scientists looked at a very long bit of DNA, and made artificial words by breaking up the DNA into 'words' each 3 rungs long. And then they tried it again by making 'words' 4 rungs long, 5 rungs long, and so on up to 8 rungs long. They then analysed all these 'words', and to their surprise, they got the same sort of Zipf Law/straight-line graph for the *human DNA* (which is about 97 per cent introns or 'junk') as they did for the *human languages*!

So at least one test (Zipf's Law) implied that there was some order in the introns. The scientists then tried another test – the test of 'redundancy'.

It turns out that one thing common to all human languages is the concept of 'redundancy'. You can drop letters, or sometimes even entire words from a book, and yet you can still read and understand the text. When they applied the mathematical tests of redundancy that are used on human languages to the introns, they found that the 'junk DNA' passed the test of a language! It could be just a coincidence.

Mileposts

In 1992, the British science journal *Nature* reported that 'the first complete sequence analysis of an entire chromosome from any organism' had been completed. This was the third chromosome (out of 16 chromosomes) of bakers' yeast, *Saccharomyces cerevisiae*. There were some 315 000 rungs mapped!

In 1995, Craig Venter, director of the Institute for Genomic Research, announced that his team had succeeded in being the first to map the entire genome of a free-living creature. In fact, the team had done it twice. They had mapped the DNA of two bacteria, *Haemophilus influenzae* and *Mycoplasma genitalium* (involved in infections of the reproductive tract).

In mid-1995, the American science journal *Science*, reported that the entire 14 million rungs of bakers' yeast, *Saccharomyces cerevisiae*, would be mapped by the end of 1995! The technology had sped up enormously in the three years since 1992.

Introns and Other 'Junk' DNA Aren't 'Junk'?

Introns are one member of the big family we call 'junk DNA'. But there are a few other types of 'junk DNA'. These go under obscure names such as Satellites, Minisatellites, Microsatellites, 3-UnTranslated Regions (3-UTRs), Heterogeneous Nuclear RNA (hnRNA), Short Interspersed Elements (SINEs), Long Interspersed Elements (LINEs) and Pseudogenes.

There is at least one clue that introns have some use. Consider the DNA that 'makes' part of the immune system. Now compare this DNA in a human being with the equivalent DNA in a mouse. In each case, well over 90 per cent of the human DNA and the mouse DNA is 'junk' – in other words, it has no known function. But over 70 per cent of the human and the mouse DNA is identical – even though it is supposedly junk!

Why should it be identical if it's just junk? If Mother Nature goes to the trouble to save the information in them accurately, there *might* be something useful about the introns.

Changing Humans?

When a car comes out of a factory, it has no knowledge of who made it or how it was made, and it certainly can't change itself. But for the first time in the history of the human race, we are getting to understand how we are made – and so we might be able to change ourselves!

One Nobel Prize winning physicist claimed that the best shape for a human being would be a cloud of iron vapour, weighing about 60 kilograms, and about the size of our planet! Now he was just letting his imagination roam free, but he said that this shape had many advantages. You could travel through space without any external machinery or propulsion; you could still communicate with your fellow humans even if they had the traditional two arms and legs; and best of all, although you could still vote, nobody could make you pay taxes!

Rungs of the DNA Ladder of Life

DNA, deoxyribo*nucleic* acid, uses four different chemicals, belonging to the family called *nucle*otides. The four nucleotides are adenine (usually shortened to A), cytosine (C), guanine (G) and thymine (T). A always pairs up with T, and C always pairs up with G, so each rung can be either A-T, T-A, C-G, or G-C. By convention, each of the rungs is always looked at from the same side, and so its name is shortened to A, T, C or G. Because the rungs are about the same length, the sides of the DNA ladder are not too bumpy.

But it seems to suggest there may be some sort of language buried in the so-called 'junk DNA'! Certainly, the next few years will be a very good time to make a career change into the field of genetics.

So now, in the mid-1990s, we have a reasonable understanding of the 3 per cent of the DNA that makes amino acids, proteins and ultimately, babies. And the remaining 97 per cent? Well, we're pretty sure that there is some language buried there, even if we don't yet know what it says. It might say, 'It's all a joke', or it might say, 'Don't worry, be happy', or it might say 'Have a nice day, lots of love, from your local friendly DNA!', or it may say something much more profound.

References

Nature, Vol. 357, 7 May 1992, 'The complete DNA sequence of yeast chromosome III' by S.G. Oliver et al., pp.38–46.

New Scientist, No. 1879, 26 June 1993, 'The cryptographer who took a crack at "junk" DNA' by Andy Coghlan, p.15.

Science, Vol. 263, 4 February 1994, 'Mining treasures from "junk" DNA' by Rachel Nowak, pp.608–610.

Science, Vol. 266, 25 November 1994, 'Hints of a language in junk DNA' by Faye Flam, p.1320.

The American Physical Society, 5 December 1994, 'Linguistic features of noncoding DNA sequences' by R.N. Mantegna et al, pp.3196–3172.

The Sciences, May/June 1995, 'Talking trash – "Junk DNA" may carry messages after all' by Haley Buchbinder, pp.8–9.

Science, Vol. 268, 2 June 1995, 'Venter wins sequencing race – twice' by Rachel Nowak, p.1273.

Science, Vol. 268, 16 June 1995, 'Closing in on the complete yeast genome sequence' by Nigel Williams, pp.1560–1561.

MAGNETIC BEES SING AND DANCE

TO MAKE JUST 1 kilogram of honey, honeybees have to fly a total distance equal to 10 trips around the Earth. In covering this enormous distance, the bees make a total of some 10 million separate visits to flowers. To tell each other where the best nectar and pollen is, the bees talk to each other. They also have to navigate their way back to the nest, without getting lost.

Making Honey

Honey is a variable mixture of the sugars fructose and glucose (about 75 per cent), combined with water, traces of vitamins and minerals, and some pigments from the plant from which the nectar originally came, so the source of the nectar affects the final colour and flavour of the honey. Nectar is turned into honey in a part of the digestive tract called the crop (also called the honey stomach or honey sac). This crop is roughly equivalent to our oesophagus. A foraging honeybee will suck up the liquid nectar and store it in the crop. Enzymes in the crop will start the conversion of the sucrose in nectars into glucose and fructose. Once she returns to the nest, she will disgorge the liquid into the mouth of a young worker bee, called a nurse or house bee. The nurse bee will either store it (and age it in combs) or feed other bees with it.

For a long time entomologists thought that bees 'talked' to each other by dancing. Just recently, though, one group of scientists found that dancing is not enough – honeybees have to do a song and dance! And other scientists found that to navigate home safely the bees have tiny onboard compasses!

Each year in America, honeybees produce an average of 29 kilograms of honey in each of some 5 million hives. This works out to a total of about 150 000 tonnes of honey – the weight of two really large nuclear-powered aircraft carriers. Altogether, there are some 20 000 different species of bees, ranging from 2 to 40 millimetres long. About 3000 of these species live in Australia. But only about 500 of these species of bees are social bees that live together in well-organised hives or nests. And of these only four are honeybees.

Three of these species are native to Asia. *Apis dorsata*, the giant Indian bee, makes very large nests over 1 metre across. *Apis florea* has a much smaller nest, about 10 centimetres across. The Oriental hive bee, *Apis indica*, makes a series of small nests in the hollows of trees, or in cracks in the rocks. The western honeybee, *Apis mellifera*, is found in Africa, Europe and the USA.

Apart from silkworms, honeybees are the only insects that have a major commercial benefit for us humans – and it's not just the honey they make. Honeybees are the most important insect involved in pollinating our crops – fruit crops (such as apple, grape and strawberry) and seed crops (such as broccoli, carrots and turnips). In fact, the value of the crops that the bees pollinate (about $US10 billion per year in the USA) is nearly 20 times as much as all the honey and beeswax that the bees make!

Now to make their mountains of honey, the bees have to eat a lot of nectar and pollen. Nectar is a sweet liquid rich in sucrose, which the plants provide as an irresistible lure to attract insects. The plants usually make it at the base of the petals of many flowers, in plant glands called nectaries. Pollen is the plants' equivalent of sperm – it forms part of their sex life. Grains of pollen are usually yellowish and between 25–50 microns in size (about one-third to one-half the thickness of a human hair). The bees carry it on specially branched, feathery hairs on their hind legs.

What the plants hope is that once they've had their fill of nectar, the bees will carry some pollen to another plant, and fertilise it. Of course, the world isn't perfect, and the bees eat both the nectar and the pollen – after all, the pollen is made from a delicious and nutritious mix of fats, proteins and carbohydrates. But the plants and the bees have worked out a reasonable

New Queen, Old Queen

The single queen in the nest emits a special smell (called a pheromone) that stops the worker bees from preparing a new queen bee. If this smell is not present, the ovaries of the worker bees will grow back. These special smells weaken as the queen is getting near the end of her six to nine year life span, and/or if her nest is so big that her smells don't make it to every corner of the nest. When this happens, within 48 hours the worker bees will start preparing special egg cells to raise a new batch of queen bees.

The first queen bee to hatch will kill the remaining baby queen bees. Then the previous queen bee sits outside of the nest with half of the worker bees from the hive. Her scout bees go and find a location for a new hive. When they find one, they report back to the queen bee and she flies off to her new nest, leaving behind the new queen bee in control of her old nest.

Bee Genders

There are three sexual 'orientations' of bees – queen, drone and worker. From the hatching of the egg, it takes 16 days to grow an adult queen, 21 days for an adult worker, and 24 days for a mature drone.

The queen is the only fertile female in the hive. There can be up to 60 000 females in the hive, but she is the only one with ovaries. She's bigger than the other bees, and has a sting which is smooth and unbarbed. This means that after stinging another creature, she can pull out the sting without killing herself. She can live for up to six to nine years.

Early in her life she has her only period of mating, in which she mates with up to half a dozen drones. She stores the sperm from these brief encounters for the rest of her life, and can release the sperm so that they meet and fertilise the eggs if she wishes. Her ovaries can lay eggs at the rate of one per minute, which works out to laying her own body weight in eggs each day! Most of these eggs have sperm added to them, but a few don't.

The eggs hatch in about three days. If an egg is fertilised with sperm, it turns into a worker, who like the queen is female. If it is not fertilised with sperm, the egg turns, via virgin birth, into a male drone.

There are very few drones in any hive or nest. They have no sting, and no ability to collect nectar. Even though these male drones come from an unfertilised egg, they are fertile and can manufacture sperm. Their only job is to fertilise the young queen bee. Over a few days, the queen will mate with six or so drones. New drones are 'manufactured' in the springtime, so that they can mate with the virgin queen if one has hatched. But each autumn they are kicked out from the colony to starve to death as winter approaches.

Workers are female. Even though they come from a fertilised egg, they are infertile and do not have any ovaries and so cannot make any eggs. As the name implies, they're there to work. From when they're born up to about day 3, their job is to clean the hive. By day 5, glands mature in their mouths and they become nursing bees. By day 12, their wax-making glands start working, and they're building structures in the nest and storing nectar and honey. Around days 18–20 they begin guarding the nest, and at day 21 they leave the nest and go foraging outside for nectar and pollen. They normally die around day 42.

relationship with each other over many millions of years. Enough pollen is spread around to keep baby plants being born, and enough nectar and pollen is eaten to keep the bees alive.

As honeybees have always seemed so well organised in their foraging it was obvious that they were 'telling' each other where the food was. The big question, however, was – how do they communicate with each other?

The answers to this puzzle began to emerge back in 1923 when Karl von Frisch from the University of Munich in Germany published his first paper on the language of the honeybees. He spent a lot of time looking at honeybees. By 1943, he had discovered that they 'spoke' with each other by *dancing a special dance*!

Suppose a honeybee has found a flower bed just dripping with delicious nectar and pollen. She then flies back into the hive to tell her fellow workers about the flowers. She dances two different dances, depending on whether the food is near or far.

If the flowers are within 100 metres of the hive, she flies in circles inside the hive. Soon, her fellow workers leave the hive, and spiral out in ever-enlarging circles until they find the flowers. But if the flowers are further away (up to 3 kilometres away), she dances a different dance inside the nest. She flies in a straight line, at the same time waggling her rear end; flies a curved line back to the beginning of the straight line; and then she dances the whole circuit again. It's the straight line that she dances on her waggling run that's important. If it points vertically up, then the other bees leave the nest, and fly in the direction of the Sun. And if the straight line points $60°$ to the right of vertical, then the other bees leave the

Special Uses of Honey

Honey absorbs and holds water vapour, so the baking industry uses it to keep baked foods fresh and moist. But it's also an excellent food preservative, because of its acidity and high sugar content. Honey has long been used as a sweetening agent, and to make that ancient alcoholic drink mead, which has been made in Europe for thousands of years.

Honey also has medical uses. It has been recommended as an antiseptic since 600 BC at least, when it was used in traditional Ayurvedic Indian medicine, and in World War II, the Russians used it to treat burns on soldiers.

Dr Vijay Raina from Jabalpur Medical College was familiar with these uses of honey, so he conducted a study to see how good honey was at stopping the growth of bacteria in wounds. He applied this honey, in various concentrations, to 60 open wounds. If the honey was applied at 30 per cent concentration, all the wounds became totally sterile (bacteria-free) after 6.3 days. If the concentration of honey was less than 30 per cent, bacteria would grow on the wound. Using only conventional dressings, 30 per cent of the wounds were still infected with bacteria after 10 days.

Dr Raina thinks that part of the explanation is that honey absorbs water from the wound, drawing bacteria with the water. Another possible explanation is that the viscous honey acts as a barrier, which stops bacteria from invading.

Rectal Temperature of a Bumblebee!

Only two of the 20 000 species of bees live above the Arctic Circle. The queen *Bomus polaris* bumblebee has only a very short summer in which to lay her eggs and get the colony set up for the winter. So she increases her internal temperature, and her eggs begin to incubate before she even lays them!

The non-scientific journal, *The National Enquirer,* found out about this research, and presented it in the form of an exposé. The newspaper 'exposed' scientists for wasting public money in studying 'the rectal temperature of a bumblebee'!

Busy as a Bee?

Bees aren't busy. They work for only 20 per cent of the time, and spend the rest of the time just hanging around.

nest and fly in a direction $60°$ to the right of the Sun.

The speed of her loops provides information too. It tells the other bees the distance to the flowers – the faster she completes a circuit, the closer the food!

Now this was a lovely and elegant theory, and Karl von Frisch received a Nobel Prize for it in 1973. His theory was correct as far as it went, but it didn't go far enough. Most beehives are pretty dark, and like us, honeybees can't see very well in the dark. So how can they see each other do the dance?

The answer is that the honeybees have to use their 'ears' as well. In the 1960s, other scientists (Adrian M. Wenner, now at the University of Santa Barbara, and Harald E. Esch, now at Notre Dame University) discovered that the wings of dancing honeybees vibrate at 220 beats per second – and the wings give off a sound with this frequency. Honeybees were *singing a song* with their wings! (Wingbeat rates vary from 10/second for the white butterfly, to 1000/second for the tiny midge!)

Honeybees do have a sort-of-ear on the second joint of their antennae. It seemed reasonable that bees could hear this song, but how do you prove it?

In the late 1980s, Wolfgang H. Kirchner and William F. Towne proved it with a strange robot honeybee. It was made from brass, was covered with a thin layer of beeswax, and was slightly larger than a honeybee. It could deliver samples of food (droplets of sucrose solution) to the other honeybees to taste, via a thin plastic tube protruding from its head. It had razor blades for wings (run by electromagnets), and it could dance (run by a modified computer-controlled X-Y plotter). It could *sing the song* with its razor-blade wings, and *dance the dance* with power from its electric motors.

Real honeybees would ignore this robot razor-blade honeybee if it just danced the dance *or* just sang the song. But when it did *both* the song *and* the dance, the real honeybees would listen to it, and obey it. The scientists could actually talk to the animals! They could program their robot honeybee, and get it to send the real honeybees out of the nest in any direction they wanted!

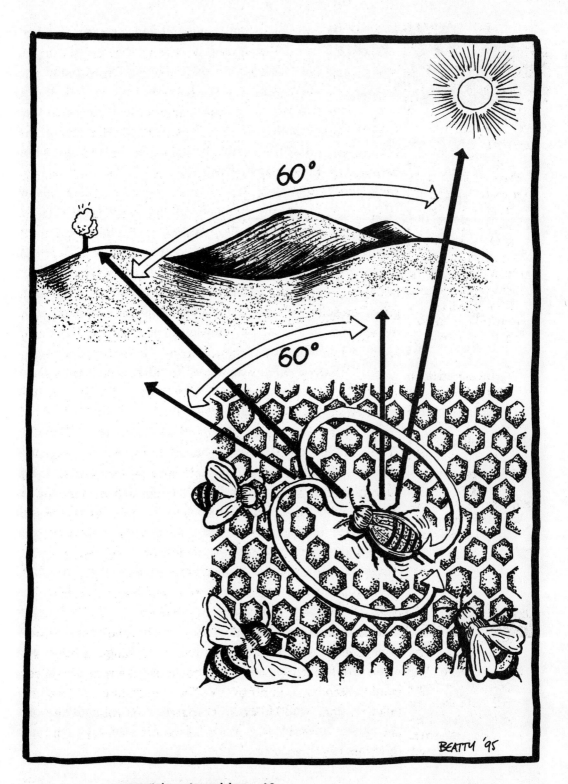

60°

60°

BEATTY '95

Ligurian Honeybees on Kangaroo Island

On Kangaroo Island in South Australia, there are signs warning tourists to beware of bees in the water taps! These bees are probably the 'purest' Ligurian bees left on the planet. In 1881, August Fiebig imported 12 hives of bees from Liguria in Italy. This strain of bee has not interbred with any other bee, since there are no other bees on Kangaroo Island. The Ligurian bees are important to the bee industry worldwide because they are a single pure genetic strain. Thanks to a quarantine, these bees have remained disease-free, even though the mainland has suffered an outbreak of foul brood disease. A wide range of honeys produced by these bees is available, from the light fragrant oilseed rape honey to the dark tangy melaleuca and banksia honeys.

So by using a song-and-dance routine, the bees can tell each other the best place to eat out. But once they've picked up their load of delicious nectar and pollen, how do they find their way back to the hive? The answer is magnets! In their tummies, honeybees have tiny compasses that sense the Earth's magnetic field – and that's how they navigate home.

Many creatures can sense magnetic fields. Under the right conditions, magnetic fields can affect humans. Susan Blackmore, a senior lecturer in psychology, wrote in the *New Scientist* about the experiences she had after a neuro-scientist at the Laurentian University of Sudbury in Ontario, Michael Persinger, had blasted her brain with intense magnetic fields in his laboratory.

She felt nothing for the first 10 minutes. Then, even though she knew that she was reclining perfectly still in a chair, she felt as though she was swaying on a hammock. Almost immediately afterwards, even though she knew that there was nobody near her, she could feel 'two hands grabbing her shoulders and pulling her upwards'. As the magnetic fields continued to act on her brain, she could 'feel' something grab one of her legs and try to pull it up the wall – although her eyes told her nothing was happening!

And then the magnetic fields began to act on her emotions. She suddenly felt very *angry* – but she didn't know what she was angry about, nor at whom she was angry. This anger lasted only 10 seconds, but as it faded, she was suddenly beset with a very intense attack of *fear*. Again, she was not scared of anyone or anything in particular, but she was very afraid.

Now, the human brain is very complicated, and we don't know why intense magnetic fields can cause such dramatic changes. But we do have a better idea of what's going on in

Royal Jelly

Right from the beginning, eggs which are destined to turn into queen bees are placed inside specially made cells which stick to the ceiling. They are stuck there with royal jelly, which is sticky, like a paste. These cells are bulbous in shape, unlike the cells of the worker bees, which are much smaller, hexagonal, and arranged horizontally.

The food for both the worker bee babies and the queen bee babies comes from glands in the nursing bees. A *clear* liquid comes from the hypopharyngeal gland, while a *whitish* fluid comes from her mandibular gland. It's the white component that seems to make a queen bee.

Royal jelly is a 50:50 mix of these clear and white fluids. It contains very large amounts of RNA and DNA, all the common amino acids, as well as cholesterol, vitamins, sugars, proteins and water. The queen bee is fed for the first days of her life on this royal jelly. In fact, she is given too much to eat, so that by the fourth day of her life she is swimming in a sea of her food – royal jelly.

But the worker bees are fed only 30 per cent of their diet as white fluid for the first 2 days of their life. This drops down to 15 per cent by day 3, and by day 4, there is no white fluid at all in their diet. This absence of the white fluid seems to stunt their ovaries so they don't grow. It also stimulates the formation of pollen baskets on their legs, which they use to carry pollen back to the nest.

Some alternative health practitioners use royal jelly on humans, but the ➤

honeybees.

There are a few different types of magnetic materials. One is a type of iron oxide called magnetite, which is naturally magnetic. We know that lots of creatures (from bacteria to whales and from pigeons to salmon) have tiny magnets of magnetite in their bodies. Magnetite was finally discovered in human brains in mid-1992 by Joseph Kirschvink, a geobiologist at the California Institute of Technology.

But there's another type of iron oxide which is paramagnetic. Paramagnetic materials are themselves not magnetic, but they are attracted by magnetic fields. So a non-magnetic paper clip made of soft iron is actually paramagnetic, because it can be pulled by a magnet.

According to doctors Hsu and Li of the National Tsing Hua University in Taiwan, honeybees have tiny paramagnetic particles in their bodies. These paramagnetic particles are inside cells inside the bees' tummy. Depending on whether they are lined up side-by-side or end-to-end, these paramagnetic particles can, as the external magnetic field changes, swell or shrink. But these paramagnetic particles are attached to the walls of the cells they are in, so if the paramagnetic particles change shape, so do the cell walls. And nerves that are attached to the outside of these cells carry signals up to the honeybee's brain.

This is the first time scientists have actually followed the 'line of information' in a creature, from the magnets to its brain. The magnetic cells in the bee's tummy are a bit like tiny onboard compasses. As the bees change their orientation relative to the Earth's magnetic field, electrical signals will be sent up along nerves to the brain.

The honeybees probably get back safely to the

hive with their load of nectar and pollen by ▷ evidence for its usefulness is not sensing how the Earth's magnetic field interacts very strong. In fact, there have been a with their paramagnetic iron particles. (Of few cases of death in children, due to allergic reactions to royal jelly. course, honeybees are excellent at navigating by the Sun, but this magnetic navigation is a great back-up system. It would certainly be very useful when the Sun was covered by clouds or storms.)

So the buzzing of the bees is not simply to remind us that spring is in the air. For the honeybees, their song and dance routine is more than just entertainment – it's a live map. And a little magnetic navigation helps them avoid getting lost on the way home!

References
Scientific American, June 1994, 'The sensory basis of the honeybee's dance language' By Wolfgang H. Kirchner and William F. Towne, pp.52–58.
Science, Vol. 265, 1 July 1994, 'Magnetoreception in honeybees' by Chin-Yuan Hsu and Chia-Wei Li, pp.95–97.
New Scientist, No. 1952, 19 November 1994, 'Alien abduction – the inside story' by Susan Blackmore, pp.29–31.
Australian Doctor Weekly, 10 February 1995, 'Honey is effective as an antiseptic' by Susan Mulley, p.57.
Earth, February 1995, 'Early attractions' by Dan Grossman and Seth Shulman, pp.34–41.

MAGNETIC SEA TURTLES

SEA TURTLES HAVE been around for hundreds of millions of years – much longer than the 50 million years that mammals have been on the scene, or the 3 million years we humans have been around. Part of the reason that sea turtles have survived so long is that they have extraordinary navigational skills. These skills let them use feeding grounds which are a long way away from their nesting sites. It's only in the last few years, thanks to a recycled fibreglass satellite dish, that we've begun to understand how sea turtles navigate.

There are about seven different species of sea turtles – the leatherback, the green sea turtle, the loggerhead, the hawksbill, the flatback turtle, the Indo-Pacific Ridley and the Atlantic Ridley. Sea turtles tend to be largish animals, and can excrete waste salt from the lacrimal glands near their eyes – which explains why sea turtles seem to cry! They range across the seas as far north as Scandinavia and right across the equator down to Australia.

The most recent research on sea turtle navigation was done with loggerheads. When they emerge from their nests in the sand of the beaches of Florida, loggerhead sea turtle hatchlings have quite a journey ahead of them. First they have to make it safely off the beach of their birth into the waters of the Atlantic Ocean. Then they have to find the Gulf Stream, and ride it first north and then east, towards Portugal. Then they spend several years at sea before the females return to the beach where they were born, to lay their eggs.

But there's one very dangerous location out in the North

Atlantic. As the Gulf Stream gets closer to Europe, it splits into two branches. The northward branch would carry them to a frigid death in the cold waters off England. But somehow they manage to ignore the northward branch, and take the southward branch into warmer waters to feed off the creatures that live in the Sargasso Sea. In fact, they continue to circle around the Sargasso Sea for up to seven years.

When they're sexually mature, they somehow return to the very beach they were hatched on, to nest and keep the cycle going. The pregnant sea turtles will come onto the beach at night, and pick their spot above the high-water level. Over a few hours, they dig a hole in the sand, lay their eggs, cover them with sand, and head back to the ocean.

How do they find their way back to just one tiny beach? We do know that migratory animals use all sorts of clues to navigate. Some use polarised light from the Sun, so they can navigate in a white-out (where, because of weather conditions, everywhere you look is white and there is no shadow or horizon visible) when they can't see the Sun directly. Other clues are the position of the Sun or the stars, the direction of the wind, infra-sound (low-frequency sound such as waves beating onto a beach every few seconds), smells, and even the Earth's magnetic field.

In fact, the magnetic field of our planet is a very good navigational aid. It's always the same – regardless of the weather, or the time of day. We already know that many different migratory birds and fish (such as pigeons, salmon, whales, tuna, sharks) seem to follow magnetic fields. In fact, scientists have found tiny little crystals of a natural magnetic material, a type of iron oxide called magnetite, in their brains. And in 1985, this same magnetite was found in the brains of loggerhead sea turtles. This natural magnet in the turtles' brain was a strong hint, but the presence

Turtle/Tortoise – Oldest Reptile?

It's often said that the turtle is the oldest reptile. Unfortunately, this is an another example of confusing language.

The tortoise/turtle that we have today is not the same as the tortoise/turtle that lived many millions of years ago in the past. For example, some of the fossil turtles have teeth, while all modern turtles are toothless – in some cases, the edges of the jaws have toothlike serrations, but they are definitely not teeth. The herpetologists prefer to talk about lineage or line of descent. So the tortoise lineage is about as old as the crocodile lineage, which is about as old as the snake lineage – and they all go back about 200 million years.

Unfortunately the fossil record is still a little incomplete, and it would be foolish for us to put our neck out, and try to claim that one lineage is older than any other lineage.

Do Turtles Return to the Same Beach?

Because there is only one species of Ridley turtle (which makes tracking the turtles easier) it has been possible to observe that Ridley turtles seem to return to lay their eggs on just one beach in Rancho Nuevo, on the Gulf Coast of Mexico.

Folk wisdom has always claimed that turtles will return to the very beach that they were hatched on, and no other, but scientific proof has always been lacking. Sure, scientists could actually do the experiment of marking each hatchling at birth, and then coming back after seven years when the turtles returned, to have a look at their shells. But only one out of every thousand hatchlings survives its seven years at sea to come back to lay eggs on the beach at which it was born. You'd have to mark a lot of turtles to do a decent experiment.

In 1990, one group of scientist decided to check it out by looking at a type of DNA that is passed directly from one female only to her daughter, and to her daughter, and so on. Anne Meyland of the American Museum of Natural History in New York worked with Brian W. Bowen and John C. Avise of the University of Georgia in Athens. They examined female green turtles from four different beaches and compared their DNA. They found that samples of the DNA taken from different breeding populations from three of these beaches were quite different from each other. This seems to indicate that these three populations were quite separate from each other.

of this magnetic material in their brains doesn't mean that they actually use it. The secret of the sea turtles' navigation was finally solved by Kenneth Lohmann, a marine biologist at the University of North Carolina, and Michael Salmon and Janette Wyneken of Florida Atlantic University.

Their work on loggerhead turtle hatchlings show there are three separate stages before the hatchlings get out into the Sargasso Sea. In each stage, they use a different navigational cue.

First, they have to move down the beach into the water (they use light). Second, once in the water, they have to head out to sea to get into deep waters (they use the direction of the waves). Third, once they're well out to sea, they have to avoid the treacherous northward branch that will lead to a frosty death in cold waters (they use their magnetic sense).

In this first stage, when the hatchlings scramble out of the sand, the clue that gets them headed for the water seems to be light bouncing off the ocean. This can be from the Moon, or from the Sun, just before dawn. In general, there'll be more light over the ocean than over the land. For these loggerhead hatchlings on Florida beaches, the direction that they have to head in is east. In fact, this initial exposure to light seems to 'set' their internal magnetic compass. If they are exposed to bright lights from the west (such as from a well-lit hotel), they will head west – and die.

The second stage happens once they're in the water. The clue that tells them where to go is the direction that the waves come from. The hatchlings seem to have a circuit hard-wired in their brains that tells them to head directly into oncoming waves. It turns out that in that part of the Atlantic Ocean, the waves practically always come directly from the ocean onto the Florida beaches.

In fact, Lohmann and his colleagues were able to test this head-into-the-waves theory when hurricane Hugo swept along the Florida coast in 1989. For a few days, the hurricane caused an unusually strong wind to blow out to sea, away from land. So they loaded their hatchlings into an esky, and rapidly took their research boat 8 kilometres out to sea. And there, as they had hoped, they found an unusual sight – the waves moving away from the land, and out to the open sea. They released the hatchlings one by one, and practically all of them swam towards the waves, even though this led them back to the shore, not out into the ocean, which was their true and ultimate home.

Wave direction might be a good navigational cue close to shore, but it's useless in deep waters. Out on the high seas, waves can come from any direction. This is why scientists thought that sea turtles used another sense once they were out on the high seas.

The third stage theory (the magnetic theory) was fairly easy to test. They got hold of an old fibreglass satellite dish, laid it on its back, and filled it with water. This was to be their artificial ocean. Over the centre of the dish they hung a hook, which they attached to a tiny Lycra swimsuit they had made for the baby turtles. This meant that the hatchling, the baby sea turtle, could swim in the same direction for ages without actually getting anywhere. Finally, they installed coils of wire (called Rubens Cube Coils) around the fibreglass dish so that they could put an

Turtle Facts

Altogether there are about 335 different species of turtles. One method of classification partitions them into two main groups, depending on how they hide their heads in the shell – the turtles that telescope their heads straight back into the shell, and the turtles that bend their necks to the side.

Turtles range in size from 7.5 centimetres (striped mud turtle, *Kinosternon bauni*) to 2.4 metres (giant leatherback, *Dermochelys coriacea*). One extinct sea turtle, *Archelon ischyros*, was as long as 6 metres! Turtles can live for a long time, sometimes for over a century. Only a few species are strict meat-eaters or strict plant-eaters. Most species have a mixed diet, but in general, the land turtles prefer plants, while the water-going species prefer meat.

All turtles lay eggs, which they bury in specially dug holes. Female turtles of some species (such as the diamondback terrapin, *Malaclemys terrapin*, and the eastern box turtle, *Terrapene carolina*) can store sperm for up to four years. Whenever they wish, they can bring the sperm into contact with their eggs. This means that after just one mating, they can release fertilised eggs for several years. Depending on the species of turtle, the eggs range in size from 1–2 centimetres up to 6 centimetres. The number of eggs in a clutch ranges from 1–100 eggs, while the incubation period varies between 45–100 days.

Sargasso Sea – Eels and Marooned Ships

The Sargasso Sea is the sluggish centre of a giant whirlpool (or gyre), created by the North Equatorial Current to the south, and the Gulf Stream running clockwise to its west and north. This seaweed-filled sea lies in the North Atlantic Ocean between the West Indies and the Azores, and between 20-35° North latitude. The Sargasso Sea gets its name from the brown seaweed that grows there – Sargassum. This seaweed gives shelter and food to many creatures, and creates a rich feeding ground for sea turtles.

Now you might not know this, but in the 18th and 19th centuries eel fanciers were puzzled by a great riddle – why were no baby freshwater eels ever found in Europe or America?

This mystery was finally solved only in the early 20th century by the Danish biologist Johannes Schmidt. The Sargasso Sea is the answer. Some freshwater eels (such as the European eel, *Anguilla anguilla,* and the American eel, *Anguilla rostrata*) will migrate to the Sargasso once they have reached maturity. Not only will they swim down rivers and creeks, they will even slither overland across moist dew-laden grass, in their need to reach the Sargasso Sea. Once there, they mate, and the female will 'spawn' her eggs, up to 20 million of them! Then she dies. The eggs turn into baby eels, which grow up at sea living on and amongst the seaweed, and then return to freshwater.

The seaweed in the Sargasso Sea is also reputed to have played another, very different role in the past. The Sargasso Sea is famous in mythology for its images of fleets of derelict sailing ships, manned by bleached white skeletons, which have been trapped in dense mats of clinging seaweed (also called gulfweed). Unfortunately for the romantics, it's not so.

There are those whirlpools that have a strong rotatory motion, like the Charybdis in the Strait of Messina between Sicily and mainland Italy, but the Sargasso Sea rotates very slowly. Maybe the legend of the marooned ships started because the Sargasso Sea is in the 'Horse Latitudes', where there is not much wind. Sailing ships in the Sargasso Sea have, in the past, been becalmed and swept by the slow current onto shores. But the seaweed is not thick, and ships can sail freely through it.

artificial magnetic field around their artificial ocean.

Lohmann and his colleagues found that once the little hatchlings were swimming happily in the upturned fibreglass satellite dish, they would usually head north-east. But when the biologists switched on the power to the wires running around the satellite dish, and reversed the apparent magnetic field, then the hatchlings immediately did a U-turn, and started heading off in what they thought was north-east, but which was really south-west. This seemed pretty good proof that the sea turtles used some sort of internal magnetic compass to navigate.

But by itself a compass is useless, unless you have a map as well. If you were dumped in a forest with just a compass, you'd be in big trouble. But if you had a map as well as a compass, you could navigate your way out, once you found a few landmarks.

Lohmann and his co-workers discovered that not only did the sea turtles have a magnetic compass, they also had some kind of magnetic map in their brain. He found out how, once in the Gulf Stream, the sea turtles managed to take the right fork, which led them back into warmer waters, and ignored the left fork, which led to the chilly waters off England.

It's all to do with the lines of magnetic force that run from Pole to Pole. You might remember from experiments at school with magnets and iron filings how the magnetic fields are parallel to the bar magnet at its middle, and are at 90° to the bar magnet at its North and South poles. The same thing happens on the Earth. At the Equator, the magnetic field is parallel to the surface of the Earth. At the Poles, the magnetic field dives straight into the ground at 90°. And in between the Equator and the Poles, it dives into the ground at angles between 0° and 90°. At the Florida beaches the magnetic lines of force dive into the surface at 57°, but at the fork in the Gulf Stream the angle is 60°.

So Lohmann and his colleagues modified their fibreglass dish apparatus so that they could actually mimic the magnetic field as it was at different parts of the Earth. And with the little hatchlings, he found that when the magnetic field dived into the water at 57°, the turtles would swim east (towards Africa). But when he increased it by adding only 3° to bring it up to 60°, the turtles would suddenly chuck a righty and head south (in the ocean that would put them in the loop that swirls around the Sargasso Sea, and away from the chilly waters of the North Atlantic). So it seems that not only do the sea turtles have a magnetic compass in their head, they also seem to have a magnetic map of important parts of the ocean.

How did they get this magnetic map in their brains? At this stage, we simply don't know.

How do they survive when the Earth's magnetic field reverses? We know that it has reversed at least 23 times in the last 5 million years. Again, we simply don't know yet.

But we do know that other sea turtles have amazing navigational skills. For example, adult green turtles feed off the coast of Brazil, but their nesting grounds are on Ascension Island. Ascension Island is about 3000 kilometres out to sea,

Poisonous Sea Turtle Soup

Late in the last century, sea turtles almost became extinct in the waters of northern Australia. The shells were turned into hair combs and spectacle rims, while the flesh was boiled into soup, which was immediately canned and exported. Green turtle soup rapidly became a fixture on European menus. As the hunters got more efficient, things looked worse for the turtles. Luckily for the turtles, though, some batches of green turtle soup were poisonous – and what was even better for the turtles, nobody could predict which batch was safe and which batch was poisonous!

Today we know that the poison is a chemical called chelonitoxin. We think that it comes from something in the sea turtles' diet, because the toxin tends to appear only at certain times of the year. This toxin is tasteless, and survives boiling (the soup is made by boiling) and canning. One unusual effect is that it will sometimes not affect a breast-feeding mother, but can kill her baby. This was known to the local Aborigines, because the naturalist MacGillivray, who explored northern Queensland in 1849 in the ship HMS *Fly*, wrote that breast-feeding mothers were not allowed to eat hawksbill turtles.

The timing of the effects is quite variable. Within hours, or a week, of a meal of turtle, the symptoms of nausea, vomiting and diarrhoea can begin. This can progress to a dry mouth, lethargy, coma and death. The death rate is higher in children, in whom it can sometimes reach 25 per cent, than in adults. The toxin causes damage (which is obvious at autopsy) to the bowel, kidneys and liver. The survivors often suffer large mouth ulcers which can take months to heal.

and this tiny spot of land (less than 90 square kilometres) is so hard to find that during World War II, air force pilots based there had a saying – 'If you miss Ascension, your wife gets a pension'.

In this case, perhaps the sea turtles use another sense to help their magnetic sense. Arthur L. Koch and his co-workers at the University of Florida have calculated that chemicals entering the water at Ascension Island would dilute by only 100-1000 times by the time they reached Brazil. Perhaps the sea turtles could 'smell' chemicals in the water, thousands of kilometres downstream of Ascension Island.

Now we humans also have little lumps of magnetite in our brains, but, because of things like clothes, make-up, under-arm deodorant, digital watches and TV, we're a long way away from our natural unspoilt original self. This could be why we

Turtle, Tortoise or Terrapin?

The difference between a tortoise and a turtle depends upon what part of the Earth's surface you're living on. In general, in Australia tortoises hang out in the water and have flippers, while turtles have feet. One herpetologist, Allen Greer from the Australian Museum, said that naming these animals was very confusing, and to simplify things, he preferred to call these creatures things-with-shells.

In the USA a turtle is basically anything-with-a-shell. If it lives in the sea it's called a sea turtle. There is also a fairly common small thing-with-a-shell that lives on land which they call a box turtle. However, there is also a fairly large thing-with-a-shell that lives in the desert and that is called a desert tortoise. On the other hand, there is also another thing-with-a-shell called a terrapin, which can either live on land or in brackish water.

In Australia we have a thing-with-a-shell which people commonly call a tortoise. This usually lives in fresh water, and not on land. On the other hand, there is another thing-with-a-shell called a turtle. It lives in the sea and has flippers.

Things are a little more confusing when you go to the United Kingdom. A thing-with-a-shell that lives on land and has feet is called a tortoise, and a thing-with-a-shell that goes to sea and has flippers is called turtle. And again there is a thing called a terrapin, but this lives in either fresh or brackish water.

can't use our magnetic compass any more. Maybe we could learn something from the sea turtles, who, like us, are armoured on the outside.

The sea turtles haven't forgotten how to use the heavy metal inside their heads, and so they don't get lost at sea. They go on to complete their life cycle, by returning years later to the very place they began.

References

New Scientist, No. 1722, 23 June 1990, 'Turtles return to "birth beach" to lay their eggs', p.14.

Scientific American, August 1990, 'The sea turtle's tale', pp.13–14.

Scientific American, January 1992, 'How sea turtles navigate' by Kenneth J. Lohmann, pp.82–88.

Science, Vol. 264, 29 April 1994, 'Sea turtles master migration with magnetic memories' by Lisa Seachrist, pp.661–662.

Earth, February 1995, 'Earthly attractions' by Dan Grossman and Seth Shulman, pp.34–41.

Australian Doctor Weekly, 30 June 1995, 'To defy turtle ban is to risk getting in the soup' by Professor Struan Sutherland, pp.35–37.

RAIN DOESN'T FOLLOW THE PLOUGH

THERE'S ONE OLD Australian saying which is heard more often in the bush than in the cities. It goes: 'The rain will follow the plough'. What this motto means is that if you come onto some land, chop down the trees, plough up the soil and plant some crops – then the rain will come. Nobody really knows if it's true, but people in the outback have often argued about it. Just recently, we've found one case where it's definitely not true – and we found out thanks to that great Australian pest, the rabbit!

There are about 300 million rabbits in Australia, and if you lined them up end to end, they would stretch 90 000 kilometres – more than twice around the Equator. Rabbits have ravished over 4 million square kilometres of Australia.

In the wild, rabbits range from 25 centimetres long and 400 grams in weight, right up to 53 centimetres and 2.7 kilograms. But domestic rabbits have been bred much larger – the Flemish giant rabbit can weigh more than 7 kilograms!

The ancient Romans were among the first to tame the wild bunny. Rabbits had arrived in Britain by the 12th century, and reached South America in the 16th century. Sailors deliberately left rabbits on South Pacific islands so that they could be used as a potential food supply for any future ship-wrecked sailors (shame about what they did to the island's vegetation).

Rabbits arrived in Australia with the convicts on the First Fleet – but they didn't really get well established at that time. Most of the 'credit' for our Australian rabbits should go to Thomas Austin, who lived near Geelong in Victoria. Before coming to Australia, he had always spent his weekends in England shooting rabbits, and being a man of strict habits, he couldn't live happily without his weekend quota of dead rabbits. So he asked his brother back in England to send him 24 grey rabbits, 5 hares, 72 partridges and some sparrows – and they arrived in 1859.

Thomas Austin was a generous man. He gave away rabbits to anyone who asked – he would even mail them. He succeeded beyond his wildest dreams in establishing a population of rabbits. But he didn't realise what his hobby was doing to the countryside. Five bunnies ate as much grass as one sheep, but one sheep was a lot more useful than five rabbits. There was much more meat in one sheep than in five bunnies, and sheep were much easier to catch than bunnies.

Thomas Austin was now happy. He had enough rabbits to kill all week long, not just on the weekends. He was so good at killing rabbits that every month, his tally of dead rabbits was reported in Victorian newspapers. The year 1867 was a good one for him – he killed 14 362 rabbits, and that was just on his property! (That's around 25–30 tonnes of dead rabbits.) By 1885, 8000 square kilometres of Victorian farm land was abandoned – because the rabbits had stripped it bare. It took only another 20 years for the bunnies to infest New South Wales and Queensland.

Soon all Australians (citizens, scientists, farmers and government officials) realised how bad the bunnies were. They tried everything to beat the rabbits, including anti-rabbit fences. In 1907, the West Australian government finished the longest anti-rabbit fence ever built – 1833 kilometres long. It ran between Cape Keraudren in the north and Starvation Boat Harbour in the south. But all the fences were doomed to failure from the very start. There were so many problems involved. Storms and drifting sand damaged and covered the

Rabbit, Rat and Hare

Rabbits were once thought to belong to the rat family. But there are some important differences between them. Rabbits have two pairs of upper incisors (front teeth) – a large pair in front and a small pair immediately behind. Rodents have only one pair of upper incisors.

Rabbits look like hares, but there are major differences at birth. Almost immediately after they are born, baby hares open their eyes and begin hopping around with their furry bodies. But baby rabbits are without any fur and are blind, helpless and quite unable to fend for themselves.

It gets even more confusing when you look at popular names. The American jackrabbit is actually a hare, while the Belgian hare is actually a rabbit!

French Who Love Rabbits

In Australia, the CSIRO has a long history of unleashing myxomatosis on wild bunnies. But in the Hérault region in south-west France, a research team from Toulouse University is trying to vaccinate rabbits against myxomatosis. They've found that if they soak fleas in a solution of the vaccine for just 30 seconds, the fleas absorb a lot of the vaccine. For the next few days, the fleas can then pass this vaccine onto the rabbits whenever they bite them.

This project was started and funded by hunters who were angry at the shortage of cute little bunnies to shoot and kill. The hunters must have been serious shooters (and rich). Not only did they pay the university about $30 000 to do the research, they are now building their own research laboratory (with a $250 000 budget) with the aim of releasing large quantities of vaccine-soaked fleas. And what will they do with their flea-vaccinated immunised bunnies? They'll shoot them dead!

fences. Kangaroos, emus and wombats made holes in the fences – the holes were hard to find and fix because of the huge distances that the fences covered. And by the time the fences were finished, the rabbits had already passed most of them.

Fences didn't work, and neither did guns, so it was time for a new approach. In the early part of the 20th century, scientists suggested infecting the rabbits with a disease that would kill them, but not harm other animals. One such disease was myxomatosis. Myxomatosis comes from the myxoma virus, a virus that's closely related to small pox and chicken pox. 'Osis' is a medical word meaning 'state of', so 'myxomatosis' means 'the state of being infected with the myxoma virus'. (For some unknown reason, hungry rabbits survive myxomatosis better than well-fed ones.)

Myxamatosis killed rabbits in South America. In 1919, a Brazilian virologist suggested to the Australian government that they use the myxoma virus to solve our rabbit problem. He received an official reply, which refused his offer: 'The trade in rabbits, both fresh and frozen, either for local food or for export, has grown to be one of great importance, and popular sentiment here is opposed to the extermination of the rabbit by the use of some virulent organism'.

But the rabbits got worse, so in 1924, the director of Veterinary Research imported some myxoma virus from Brazil into New South Wales. Laboratory tests showed that the virus killed rabbits if they were directly injected with the virus. But the virus didn't really spread easily from sick rabbits to healthy rabbits. Farmers couldn't go out with a syringe and personally inject every healthy rabbit in Australia, so interest in the myxoma virus dropped. Today we know that

British Who Love Rabbits

Ecologists at the University of East Anglia in England have been studying Breckland Heath, a 940 square kilometre site in the south-east of England. It was one of the first areas to be colonised by early hunter-gatherers in England some 5000 years ago, when they came across from Europe. They brought grazing animals with them — which helped turn the original open woodland into its current bleak state. This heath is now quite desolate, with sparse vegetation and many sandy areas. There are several different types of soils there, but most are fairly infertile. Even so, they say it is an important conservation site for rare plants, birds and insects.

Much of this heath has been heavily grazed by rabbits over the last 700 years. The rabbits were closely guarded (from the working class), because of their valuable meat and fur. In 1813, one rabbit poacher was sentenced to seven years' transportation for stealing a single rabbit. So valued were the rabbits that their natural predators were killed and the rabbits were even provided with meals-on-wheels during harsh winters or droughts.

These ecologists, Paul Dolman and William Sutherland, want to keep things the way they were — nice and bleak. They reckon that rabbits preserved the bleakness of the heath, by grazing and turning over the soil. They wept over how myxomatosis reduced the rabbit population in the mid-1950s, and how the heath rapidly became overgrown. The annuals and bare ground were replaced by perennials and rank grasses. But then they say in their *New Scientist* article: 'thankfully, the hot summers of the past two years have helped to restore populations (of rabbits) in some areas...' — in other words, to bring back the bare ground!

Dolmen and Sutherland are working with the Nature Conservancy Council and the Norfolk Naturalists Trust to find ways of reducing the fertility of the soil, such as stripping off the top soil, ploughing and forage harvesting — because the rabbits are taking too long to devastate the landscape.

It makes an odd sort of sense. The Breckland Heath has always been interfered with by ancient flint workers and farmers, the military, the rabbits and the rabbit warreners. If they want to keep their heath in a semi-desert state without having to use bulldozers, then they'll need more rabbits.

Americans Who Love Rabbits

A Massachusetts company, Transgenic Sciences, wants to turn bunnies into 'commercial bioreactors'. Genetic engineers have already modified bacteria to make chemicals that we humans find useful — chemicals like insulin, human growth hormone, blood-clotting agents like tissue plasminogen activator and the chemical missing in some haemophiliacs, factor IX. Transgenic Sciences wants to genetically engineer rabbits so that the modified rabbits will make chemicals we can use — but at one-third the cost of bacteria. They chose rabbits because rabbit milk is already quite rich in proteins and could easily carry a few extra chemicals.

Japanese Who Don't Love Rabbits

The rabbits are getting the last laugh in Japan. In a survey of 5600 laboratory staff who work with various laboratory animals, 30 per cent of them were 'allergic' to rabbits. They suffered from runny noses and tight breathing whenever they came near the bunnies. Will these lab workers go out and work with the animal liberationists, to get away from the bunnies?

Odd Rabbit Habit

Rabbits eat their own faeces. When there is not much feed around, the rabbits produce a different type of dropping called a 'soft faeces'. They eat it directly out of their anus and swallow it. It's a bit like cattle chewing on the cud, and this odd habit lets them get a few extra nutrients out of the food.

insects, such as fleas or mosquitoes carry the myxoma virus from one rabbit to the next. In fact, scientists call these insects 'flying pins'.

The NSW Department of Agriculture infected rabbits with the myxoma virus around 1926, but as the disease didn't really spread, the rabbit problem continued to grow steadily worse. In 1934, Dame Jean MacNamara, an expert on poliomyelitis (caused by the polio virus), suggested that the myxoma virus should again be used against the bunnies. In 1936, scientists injected rabbits with the myxoma virus in one of the drier parts of South Australia. (In fact, South Australia is the driest state in the driest inhabited continent on Earth.) As before, the virus killed the rabbits that were directly infected, but didn't spread. At that time, the scientists still hadn't realised that the myxoma virus needed insects to spread from one bunny to the next.

The rabbits were now totally out of control. In the 1940s, there were hundreds of millions of rabbits – and each one cost the Australian economy one dollar. Myxomatosis worked – but only locally, and only for a short time.

In 1949 there was yet another rabbit plague in Australia. Dame MacNamara again suggested using the myxoma virus, and wrote articles for the popular press, such as *Stock and Land*, and the Melbourne *Herald*. This time the scientists ran their experiments in the Murray River area. Luckily for them, the rain was pouring down, so there were lots of mosquitoes around. Again they injected the myxoma virus into the rabbits, and released them. For a while it seemed as though they were heading for another failure, and the dispirited scientists headed back to Canberra to write up their report.

But a long distance call from a farmer lifted their spirits. He said that there were sick rabbits all over his property. The scientists soon realised that the rain had provided breeding sites for the mosquitoes, which could then spread the virus.

By terrible coincidence, at the same time as the rabbit populations plummeted by nearly 99 per cent in the Murray valley, there was an outbreak of human encephalitis in the

AUSTRALIAN FARMER MAINTAINS THE
STATUS QUO

BEATM '95

Rabbits' Evolution in Australia

Rabbits are not only good at breeding, they're also really good at evolving. Practically all of the 300 million rabbits in Australia came from the 24 grey rabbits imported into Australia by Thomas Austin in 1859. But since then, they have rapidly evolved the shape of their ears to suit their local climate. Long floppy ears are useful in a warm climate because they radiate away the heat. But in the Alpine parts of Australia, they can actually be a liability, because they waste precious heat.

Dr Kent Williams and Mr Robert Moore of the CSIRO Division of Wildlife and Ecology found that desert rabbits had longer ears than the mountain rabbits. And when they breed the rabbits from the deserts and the mountains in different conditions in the laboratory, within a single generation the rabbits began to adapt to their new artificial environment.

So one secret advantage that rabbits have is their recently discovered ability to adapt rapidly to new climates.

same area. Several children died, and hundreds of adults became very sick. The farmers began to mutter about the scientists having released a 'killer disease'. But they soon became convinced that the myxoma virus was not the cause, when three leading Australian scientists publicly injected themselves with the live myxoma virus to show that it was not dangerous to humans.

Rabbits are still a real worry in Australia, and various scientists are trying new and different ways to fight the bunnies. But although rabbits might be very bad for the Australian ecology, environment and economy, they might also have accidentally given us a very important lesson in agriculture.

From outer space, the most visible artificial feature on the Australian continent is caused by a fence. This fence is a rabbit fence, which stretches some 250 kilometres inland from the Western Australian coast near Esperance. This 100-year-old fence didn't stop the bunnies from the eastern states heading west – but it did stop the farmers in the west from growing their crops east of the fence. But the odd thing is that from space, you can see more greenery on the bunnies' side of the fence! It's this sharp division that's the most visible artificial feature on the Australian landmass.

In that area of Western Australia, the rainfall used to be at its maximum way down in the south-west corner, and it used to drop gradually as you headed either north or east. But this pattern has reversed now that 130 000 square kilometres of native vegetation has been cleared. On the west side of the rabbit fence, where the farmers grow their crops, the rainfall has dropped by 20 per cent over the last 40 years. But on the relatively untouched eastern side of the rabbit fence, the rainfall has actually increased by 10 per cent over the same period! The farmers even say that sometimes it rains just on the other side of the fence, and avoids their farmlands!

This seems to show that, at least in this area, land clearing actually reduces the rainfall! Contrary to the old farmers' saying, the rain does *not* follow the plough – in fact, it seems that agriculture sends the rain away! It's a bit of a fluke that the conditions to test this saying all came together in the south-west corner of Western Australia.

Professor Peter Schwerdtfeger, the professor of Meteorology at the Flinders University in South Australia, thinks he has found the mechanism to explain what's going on. He used two specially equipped research planes loaded with instruments. These flying laboratories can simultaneously measure the energy coming from the Sun and the ground, the temperature of the ground, and the concentrations of various gases in the air. Not only can they measure the direction and velocity of the wind, but even its temperature and humidity.

It turns out that both rainfall and the saltiness of the soil are controlled by the temperature of the soil. It all starts with the Sun – that giant hydrogen bomb in the sky that consumes 4 million tonnes of hydrogen each second! The Sun dumps onto each square metre of ground a maximum of around 1 kilowatt of power – enough to run a small electric heater.

Some of this power is absorbed by the ground and plants, heating them up. And some is just bounced straight back into space – without heating either the ground or the atmosphere. Professor Schwerdtfeger found that the scrubland (the bunnies' area) reflected into space about half as much energy as the farmland, depending on

Rabbit Eradication ①

One single female can produce up to 25 live baby bunnies each year, if she's living in an easy climate like southern inland New South Wales. In fact, rabbits are such good breeders that if you want to keep the population steady, you need an 85 per cent death rate.

Some CSIRO scientists are trying to use this high birth rate against the bunnies. As the sperm from the male rabbit heads towards the egg in the female rabbit, it has to pass a number of 'barriers' at the vagina, the uterus, the fallopian tubes and finally at the egg. On the outside of the sperm there are special proteins that let each of these layers know that the sperm is a welcome guest. But it's possible to infect the female rabbits with a modified virus that stimulates and tricks the female rabbit's immune system. Her immune system now thinks that these 'friendly' male proteins are actually bad guys. In other words, the female rabbits can be vaccinated against becoming pregnant!

Rabbit Eradication ②

Another way to get rid of rabbits is to improve the ability of the insects (the flying pins) to deliver the myxoma virus. So other scientists have been looking for fleas that can prosper in drier areas. Other scientists are looking for more lethal forms of the myxoma virus.

the season, and the types of crops grown. All year round, the scrubland absorbed more of the Sun's energy. In other words, the ground and vegetation in the scrubland was hotter than the farmland. He also measured about two and a half times as much air moving upwards over the scrubland as compared to the wheatfields. This was because the ground was hotter, and was sending the air upwards (after all, hot air rises).

Plants suck water out of the soil, and breathe it out from their leaves. So, over the scrubland, any water sucked out of the ground by the plant life was more likely to be lifted up high enough to make rain clouds. The pastoral scrubland had all sorts of different bushes and trees – so no matter what the season was, there was fairly reasonable groundcover of some sort all year round. But the 130 000 square kilometres of farmland that had been cleared had only a few different crops, and it had good-sized plants on the surface for only a month or so each year. So because, on average, there was less groundcover, the farmland was less likely to make rain clouds.

Rabbit Eradication ③

But intensive monoculture in this area does more than keep the rain away – it also makes the soil very salty! It turns out that the plants have another role, to keep the underground water away from the surface. In much of Australia there is salty water some distance under the ground. So long as the level of this water is not near the surface, it causes no immediate problem. If plants suck up some of this water through their roots, and then pass it upwards as water vapour into the sky, then there is a constant 'drain' on the underground water reservoir and so the water level remains some distance underneath the soil.

But suppose that the plants can't pass this water up into the sky. Any water (say, pumped onto the soil from a nearby river) will enter this underground water reservoir, and it will stay there. With time, the amount of water builds up,

Yet another way to kill rabbits is quite controversial – RHD. RHD does not stand for Right Hand Drive, but for Rabbit Haemorrhagic Disease.

RHD is a nasty disease. It attacks the liver and spleen of the bunny, and kills it in a few days. Unfortunately, it doesn't kill them painlessly – in the last hours before they die, they scream in apparent pain. The scientists are not too sure where the virus originally came from, but some of our earliest records show that it appeared in China in 1984. It has spread to Europe, and has already killed some 64 million farmed rabbits in Italy alone. But even if it does work in Australia, after some years the virus will inevitably weaken and the rabbits will build up a resistance to it.

and the salty ground water gets closer to the surface. Already in that south-west corner of Western Australia, the farmers are finding that the patches of salty dead ground are getting bigger.

To summarise Professor Schwerdtfeger's results: on the unfarmed bunnies' side, the soil gets hotter, moist air is lifted into the sky to make rain clouds, and the salt water stays underground; on the farmers' side, the ground doesn't get so hot, the moist air stays close to the ground, there are fewer rain clouds, and the salt water comes to the surface. So trees are more than just pretty – they 'make' rain clouds, and keep the salty water safely underground.

Now this is just one experiment carried out in just one location over the fairly short time of half a century. So at this early stage, we really can't say that it's absolute truth – but it's pretty suggestive.

And what can we do about it? Well, nobody's done the experiment, but Professor Schwerdtfeger suggests that the farmers try replanting the land in a striped zebra pattern, with a 50:50 ratio of trees and crops. It might be expensive for the farmers to take so much valuable cropland out of production – unless they can plant economically valuable trees (like sandalwood and quondong, which are both native to Australia) and deep-rooted shrubs that could yield attractive wildflowers, for which there is a growing Asian market. The farmers will have to do something different, because they are gradually running out of options. Thanks to that 20 per cent reduction in rainfall, in some years, there's simply not enough rain to grow wheat at all.

In our attempt to stop the bunnies devastating the land, we accidentally found that clearing the

Rabbits and Akubra

We can each make our own personal contribution to the rabbit plague by buying an Akubra hat. An Akubra hat uses 8–13 rabbit skins. The Akubra I have is the Sombrero (I chose it because it was the largest hat on the market, to give me maximum protection from the Sun). Each year the Akubra company uses about 360 000 rabbit skins to make about 40 000 hats. Pressure is used to blow the rabbit skins into a metal cylinder, which is then compressed. Finally, the skins get shaped into one of the many different Akubra hat styles.

Most of the rabbits skins are genuine Aussie wild bunnies. But Akubra does import a few white rabbit pelts for the pastel-coloured hats – which are only a small part of the market.

My beloved Sombrero is about seven years old and the kids have punched a couple of ventilation holes in the top. In its pre-hole mode, we've used it to carry litres of water. It's also been a very handy standby whenever we've been in light planes or helicopters, but I'm very glad to say that I've never actually had to use it to catch the technicolour yawn. And the ventilation holes haven't made me love my hat any less – they make it cooler.

land reduces the rainfall. The bunnies have been pretty bad to Australia, but maybe they haven't been all bad. Maybe they provided us with an experiment that could bring us back to our senses, and change the way we manage our land.

References

New Scientist, No. 1712, 14 April 1990, 'Drugs industry turns animals into "bioreactors", by Susan Watts, p.14.

Ecos 64, Winter 1990, 'Ear to stay' by Roger Beckmann, p.27.

New Scientist, No. 1820, 9 May 1992, 'Fleas help hunters have their fun' by Jeremy Webb, p.4.

Ecos 71, Autumn 1992, 'New approaches to rabbit and fox control' by Carson Creagh, pp.18–24.

The Horizons Of Science Forum, organised by the Centre for Science Communication at the University of Technology, Sydney, 17 November 1993, 'The rabbit fence: a surprising climate barrier' by Professor Peter Schwerdtfeger, pp.1–8.

Time, 4 April 1994, 'Australia's most wanted' by Graeme O'Neill, pp.47–53.

New Scientist, No. 1957/58, 24/31 December 1994, 'Chinese virus to curb runaway rabbits' by Tim Thwaites, p.7.

TEA SCUM

TEA IS THE MOST popular non-alcoholic drink in the world – followed by coffee and cocoa. Tea is big business. India grows some 650 000 tonnes of the stuff each year – thanks to the invention of the portable greenhouse! And every day in the United Kingdom, people drink 150 million cuppas.

While tea has always been delicious, refreshing and invigorating, there's always been one great problem with it – the tea scum that appears on the surface of your cuppa. But now science has galloped to the rescue to find that tea scum is just chalk, or limestone – and has even provided a cure.

Tea, the drink, has an ancient history. One Chinese legend claims that in 2737 BC, the Emperor Shen-Nung became the first to brew tea when some leaves fell from a tea plant into his boiling water. One reason for the early popularity of tea was that the boiled water would be germ-free. It was apparently first cultivated in the Szechuan Province, and gradually spread to the coast. It then moved east through Java to Europe, and appeared in London in 1657, when it was first sipped at Thomas Garraway's Coffee House, in Exchange Alley.

Tea was so highly valued when it finally reached Europe in the late 1600s that it was stored in locked wooden boxes. These boxes were called 'caddies' after an Asian unit of weight, the 'catty'.

It was as tea became more popular that fervent tea-drinkers began to worry about tea scum – that thin, almost-invisible film floating on the top, which becomes brown-coloured and

Tea Money

Tea was used as official currency for 900 years in China and Tibet. Wood shavings were mixed with tea leaves and these were then pressed into a 'tea brick', weighing about 1 kilogram. If your transaction didn't come out to a round number of tea bricks, you could break a brick into smaller chunks. Tea bricks were being issued by some banks in Asia as recently as 120 years ago.

Tea and Portable Greenhouse

The event that made tea such a huge money earner for the British Empire was the invention of the greenhouse – which was itself an accident!

In 1829, Dr Nathaniel Bagshaw Ward was a doctor living in the very polluted East End of London. He loved growing plants. But the dirty air in his neighbourhood usually killed his plants.

Dr Ward was also interested in insects. One day he buried the chrysalis of a moth in a glass bottle with some moist dirt, to see how the insect would grow. Not only did the little insect grow, but so did tiny shoots of fern and grass. As the Sun heated the soil, water came out of the soil, condensed on the side of the bottle, and went back into the soil again. He realised that the plants survived because they had a stable, warm, clean environment with moist air. Provided he could add light, which travelled well through the transparent glass, his plants got everything they needed.

In 1833, he made two little portable glass greenhouses, loaded them with grasses and ferns, and sent them to Australia by ship. They survived. Some native Australian plants were then put into the same portable greenhouses and sent on a nine month trip back to England – and these plants also survived.

In the early days these miniature greenhouses were called Wardian cases. In 1848, one of these Wardian cases was used to carry tea plants from China to India – and so the ➤

can then discolour the sides of the cups, spoons and even your teeth. Finally, in 1992, the mighty engines of science were focused on tea scum when chemists Michael Spiro and Deogratius Jaganyl of the Imperial College of Science, Technology and Medicine in London were asked by a tea manufacturer to research this vexing problem.

They published their results in the prestigious British science journal *Nature*. Their letter, entitled 'What causes scum on tea?', was right at the beginning of the section called 'Scientific Correspondence'. Further down the same page was another scientific letter called 'Neurotransmission and secretion'. Presumably the placement of 'What causes scum on tea?' at the top of the page was a measure of its importance, and a mark of respect!

The chemists began their studies by heating up London mains tap water to 80°C, adding teabags, then keeping the water at 80°C and observing it – for up to four hours! They saw that over the first hour, the longer they left the teabags in, the more scum they would get.

They collected some of this scum with an aluminium scoop, for analysis. With an electron microscope, they could see an organic layer with small white patches of calcium carbonate on the surface. The scum was about 15 per cent calcium carbonate and 85 per cent complicated organic chemicals.

They found that some of the chemistry of tea scum is fairly simple. Calcium ions and bicarbonate ions will combine with each other to make calcium carbonate. They knew that this reaction could be driven backwards by an acid environment – and voila, when they added lemon juice, they got less scum!

The scum formed only if there were both calcium ions and bicarbonate ions in the water. Pure distilled water did not make tea scum, nor did pure distilled water with either calcium ions or bicarbonate ions. If they added milk (which is loaded with calcium), they got more scum than if they didn't add the milk.

So it seemed as though tea scum would form only if there were calcium and bicarbonate ions in the water. It also seemed that some of these ions solidified as little crystals of calcium carbonate – also called limestone.

But what about the organic chemicals in the scum?

Spiro and Jaganyl fiddled with the atmosphere, and noticed that they would get lots of scum when they made tea in an oxygen-rich atmosphere, and much less scum in a nitrogen atmosphere. As they wrote in their first report to *Nature*, this implied that: 'The scum was therefore produced by the oxidation of tea solubles [chemicals in the tea] such as certain polyphenolic components'. Now London tap water is a hard water loaded with minerals. So their conclusion was: 'Thus tea scum is a complex organic material derived from the oxidation of tea solubles mediated by calcium salts and accompanied by calcium carbonate.'

One of the beauties of science is that it is self-correcting. Scientists will try to repeat other scientists' work and will discuss their findings openly in scientific journals. A few months later, P.P. Jones of the Cercol Laboratories, also in the United Kingdom, began some scientific sparring. He also wrote to *Nature*. His letter, called 'Time for tea', was on the same page as other letters discussing scientific matters of equal importance

▷ enormous tea industry began. These portable greenhouses were also used to take rubber trees to Sri Lanka from Brazil, and coffee beans to Brazil from Africa.

The Japanese Tea Ceremony

Tea was supposedly introduced into Japan in 805 AD by the Buddhist priest Saicho, after a three year visit to Chinese Buddhist temples. The origin of the tea ceremony is controversial. Some say that it was introduced into Japan from China around the 12–13th century by Zen monks, while others say that the Shogun Yoshimasa introduced the tea ceremony in 1484. One Japanese teamaster, Rikye Sen, supposedly formalised it sometime before he committed seppuku (ritual suicide) in 1591.

Tea was originally brought in for its medicinal properties, and to keep the monks alert during meditation. The tea ceremony is today a very formalised ceremony, emphasising aesthetic simplicity, humility, tranquillity, peace and consideration for others. The participants traditionally enter the small room on their knees through a small door. The intricate and complex ceremony takes four hours, during which two green teas are served. Everything is exactly prescribed – the room, the surroundings, the bowls etc, and the principal guest is the first to be served by the host.

such as 'Velocities in the mantle' and 'OH in Saturn's Rings'.

He agreed that there was some kind of reaction taking place between the chemicals in the tea and the calcium and bicarbonate ions found in hard water – but he thought that this reaction was insignificant! He claimed that the brown colour in tea scum was easy to explain, and went on to give a

Tea and Iron

An advertisement for Nippon Steel in the *Smithsonian* of December 1994 extolled the benefits of iron kettles for boiling water to use in preparing tea. The author is Tomomi Kono, a director of the Kono Food Research Institute, and a professor of the Osaka Kun'ei Women's College. He has written more than 160 works.

He writes: 'An iron tea kettle is always used when preparing tea for the Japanese tea ceremony; indeed, it has long been accepted that it is nearly impossible to make good quality hot water for tea without using an iron kettle.

'Water boiled in an iron receptacle is ideal for making tea, as traces of the metal dissolve in the water and agglutinate with organic compounds, leaving hot water that is very nearly pure. More precisely, the water is partly ionized and subjected to a magnetic-like force. The minute amounts of metal dissolved in the water are drawn together, agglutinate with the organic compounds, then precipitate out and adhere to the iron receptacle. The resulting pure water helps create tea with an extremely fine flavour. A danger exists, however, that if the metal used for the receptacle is not of a very high purity, the color of the tea could be altered, or its flavour impaired.'

different explanation! He wrote that the brown colour came from the brownish tannin in the tea physically soaking into the numerous particles of calcium carbonate that were floating in the tea as a fine suspension. He didn't discuss the fact that the plates of tea scum appear solid.

Tea is so significant in most societies on our planet, it was inevitable that the 'tea scum' debate would go on. A few months later, a third party joined the scientific debate. An American scientist, Ralph A. Lewin from the Scripps Institution of Oceanography in California, wrote to *Nature*, disagreeing with both the previous reports! His letter, entitled 'Waxy tea', had pride of place next to another scientific letter on the same page entitled 'Radial single-layer nanotubes'.

He claimed that tea scum had nothing to do with calcium. He said that tea leaves were covered with a thin layer of fat, which acted as a natural waterproof raincoat and as a moisture content controller. He wrote that: 'Major components of the scum on tea are high-melting point lipids, epidermal waxes which are standard waterproofing equipment on leaves of land plants such as *Camellia (Thea)*'. He claimed that the fat just dissolved from the surface of the tea leaves in the hot water, and then floated to the surface of the water, where it cooled into flat sheets. He did agree with the originators of the discussion, when he wrote that: 'As the tea is drunk, the waxy layers stick to the sides of the cup'.

And the brown colour – well, that was nothing to do with the chemical called tannin. He wrote that a different family of chemicals called phenolics, which go brown in air, was staining this flat sheet of tea scum.

Finally, he wrote it was possible to wash the layers of scum from the cup and he described the process: 'One can liquefy, saponify and solubilise them with very hot water and detergent. (I do this every day.)'

The scientific debate soon became more involved, with a fourth comment on tea scum, entitled 'More on tea scum'. This response, the third response to the original article, was from Brazil.

In their letter, two Brazilian scientists, Enzweiler and de Oliveira, agreed wholeheartedly with the researchers Spiro and Jaganyl. They also agreed with Jones that tannin could soak into the surface of the layer of calcium carbonate, giving tea scum that famous brown colour. But they claimed that Jones' explanation of tannin soaking into floating particles of calcium carbonate did not explain the fact that tea scum floats in plates. They also disagreed with Lewin's claim that tea scum had nothing to do with calcium and was just plates of fat dissolved in the hot water from the outside of tea leaves. After all, they wrote, their soft Brazilian water had hardly any calcium, and their tea never had tea scum on it!

A few months later, and nine months after their first paper in *Nature*, Spiro and Jaganyl, with some reinforcement from another scientist, Yuen Ying Chong, published a summary on tea scum, entitled 'Scum summary'. They claimed that the brown colour of tea scum came not from tannin, but from 'insoluble oxidised [chemicals in the] tea

The Aussie Tea Ceremony = Bush Cuppa

One thing I love about going bush is making the 'centrifugal' cuppa tea. This involves a few procedures which you usually can't do in your kitchen at home.

The first procedure is as follows: Having boiled a billy of water over a fire, you then add your leaves and carefully (without spilling the hot tea and scalding anyone) swing the billy around in a circle (either vertical or horizontal). This drives practically all of the tea leaves to the bottom of the billy. You sink the remaining tea leaves by banging the side of the billy with the head of a spoon. The shock waves rush inward from the wall of the billy, and the tea leaves turn on their side, and slowly spiral to the bottom of the billy.

If there do happen to be any very stubborn tea leaves left, I then 'filter' them out using the second procedure – the special tea-pouring method. This involves holding the billy by the handle so that the thin wire handle is oriented up-and-down. With a bit of careful balancing, the billy does not lurch from side to side, and a hot stream of delicious leaf-free amber tea flows from the billy, splits around the wire handle and pours into the cup.

[called] polyphenolics'. After all, they wrote, why did the oxygen content of the atmosphere above the brewing tea have such an important effect on the amount of tea scum formed?

They too disagreed with Lewin's hypothesis that waxy lipids (or fats) from the outside of tea leaves were involved to any significant degree: 'This hypothesis does not explain why no scum forms when tea leaf is infused with hot distilled water, distilled water containing calcium chloride or soft water – or why tea scum always forms in temporary hard water, even when instant tea, which contains no insoluble waxy components, is used'.

So, in the mid-1990s, we haven't seen the last of this ground-breaking research into tea scum – but at least we do know what to do about it. Tea scum seems to be related to calcium. So if you don't use milk, and brew with soft water, you minimise your chances of getting tea scum. And because acid drives any calcium carbonate-forming reaction backwards, by adding a little lemon juice you can almost guarantee a scum-free cuppa!

English Tea Ceremony

Anna, the Duchess of Bedford, began the habit of afternoon tea in 1840. In those days, coffee was more popular than tea in the UK. This was changed by a disease, the coffee rust (*Hamileia vastatrix*). It first appeared in Ceylon (Sri Lanka) in 1869 and virtually destroyed the coffee industry in that part of the world. As the coffee prices soared, tea suddenly gained in popularity.

References
Nature, Vol. 364, 12 August 1993, 'What causes scum on tea?' by Michael Spiro and Deogratius Jaganyl, p.581.
Nature, Vol. 365, 7 October 1993, 'Time for tea' by P.P. Jones, p.498.
Nature, Vol. 366, 16 December 1993, 'Waxy tea' by Ralph A. Lewin, p.637.
Nature, Vol. 367, 17 February 1994, 'More on tea scum' by Jacinta Enzweiler and Marcelo G. de Oliveira, p.602.
Nature, Vol. 368, 28 April 1994, 'Scum summary' by Michael Spiro, Deogratius Jaganyl and Yuen Ying Chong, p.815.
Discover, September 1994, 'Tea'd off' by Jeffrey Kluger, pp.38–43.
Discover, 12 November 1994, 'Scum science' by Forrest M. Mims III, p.12.
Smithsonian, December 1994, 'Cooking with iron' by Tomomi Kono, p.22.

WALKING WOMEN

ALL IN ALL, WALKING is a pretty good way of getting around. On a smooth surface like a road, it might not be as efficient as a bicycle, but it certainly is a lot more reliable when the going gets rough. But one problem with walking is that when you carry a load, you have to burn up extra energy. Well, some tribeswomen in East Africa can carry heavy loads without burning up any extra energy – and even with all of our science, we still don't know how they do it!

One of the early reports on this topic appeared in 1986, when a team of scientists and engineers (Maloiy, Heglund, Prager, Cavagna and Taylor, from various universities in Kenya, Boston and Milan) wrote about their research in the British science journal *Nature*. While they had been travelling in East Africa, they had been surprised at the enormous loads that some African tribeswomen could carry. These women carry loads with their heads. Now you and I can carry only *15* per cent of our body weight as a load on our heads, whereas these women can carry up to *70* per cent!

Since 1986, various scientists have looked at these women, trying to work out what's going on. They've used treadmills, energy-oxygen consumption devices, and force plates. (When you step on a force plate, it registers the forces as well as the movements in three dimensions – left-right, up-down and forward-backward.) The latest work was done by Heglund, Willems, Penta and Cavagna in May 1995. They compared these women to comparison groups of male army recruits, and various other non-African adults who were not used to carrying loads with their heads. The results were astonishing.

They found that these women could carry loads of up to 20 per cent of their body weight on their heads without burning up any extra energy. But the comparison group of people, who carried loads with their hands or their backs, began to burn up extra energy as soon as they were carrying any extra load. (If members of the comparison group carried a 10 per cent extra load, they burnt up 10 per cent more energy – a 30 per cent extra load consumed 30 per cent extra energy, and so on.)

Once the East African women began carrying loads over 20 per cent of their body, they then begin to burn up more energy – but this energy would be offset by that first 'free' 20 per cent of load carrying capability. So at a given load, if the untrained volunteers were burning 70 per cent extra energy, the East African women were burning some 20 per cent less – only 50 per cent. The East African women always used less energy than the comparison groups.

How on earth do these women carry nearly 20 per cent of their body weight for 'free', without burning up a single extra calorie or kilojoule!?

In the mid-1990s, we still don't have the full answer, but we think part of the solution might be something to do with using the body as a pendulum. In a pendulum, there are two types of energy involved. **Kinetic energy** (that's energy of *motion*) is greatest when the pendulum moves *fastest*, at the bottom of its swing. **Gravitational energy** (that's energy of *height*) is greatest when the pendulum is at the *top* of its swing – when it's stopped. Once you set a pendulum going, it runs for ages, continually swapping gravitational energy for kinetic energy, and back again. It runs at nearly 100 per cent efficiency, so it hardly ever needs any extra energy fed in.

Highlands and Lowlands

In East Africa there are two main groups of women who use their heads to carry these enormous loads. The women of the Luo tribe live on the western flatlands of Kenya. Because they usually cross flat terrain, they probably don't need to look down at the ground so often – so they carry their loads balanced on their heads. But the women of the Kikuyu tribe live in the rugged central highlands of Kenya. They need to see where they place their feet. If they carry their load balanced on their head, the load would fall off when they tilted their head forward to look down at the ground – so they carry their loads in a big bag on their back, but held on by a strap looped across their forehead. The load is mostly supported by their head. As they walk, they lean forward so they can easily see where to put their feet.

Some of these women spend so much of their lives carrying heavy loads that they have a permanent groove in the bone of their skull across the forehead. This groove is often deep enough to lay your little finger in!

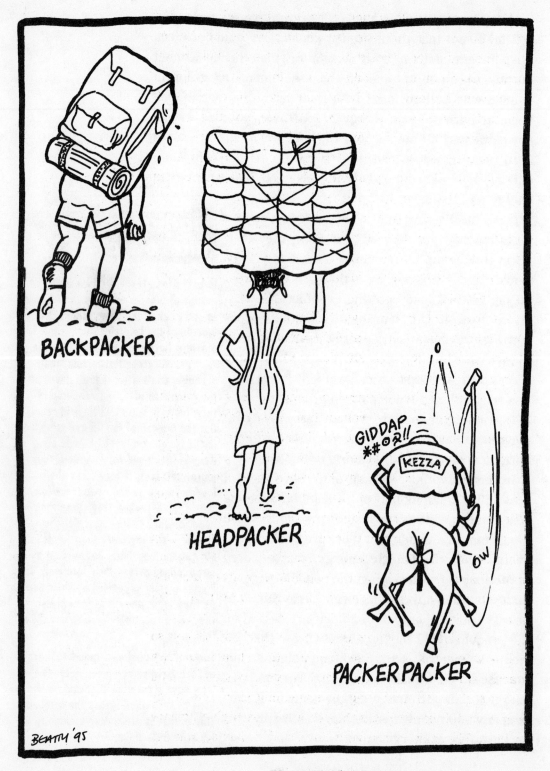

When humans walk, without carrying a load, we operate at around 65 per cent efficiency, so we continually have to feed in extra energy. But although it may seem hard to believe, we humans have a bit of 'pendulum action' in our walking too.

As you walk along, your body *rises* and *falls* during each step. It rises when your feet are side by side (and that's when you gain gravitational energy), and it falls when they are apart (and you lose that gravitational energy).

But during each step, your body also *slows down* and then *speeds up*. You actually reach your highest speed, and your greatest kinetic energy, when both of your feet are apart, and on the ground. But then as you move forward, you put your weight onto one leg, and your centre of gravity rises – but your body slows down its forward motion, and your kinetic energy actually decreases.

So during walking, when your gravitational energy is high, your kinetic energy is low – and vice versa. Your body acts like a pendulum, converting kinetic energy into gravitational energy, and back again.

Now, when they're not carrying a load, the average walking person and one of these women from East Africa will each operate around 65 per cent efficiency. When the average person carries a load, their efficiency stays the same, so they have to burn up extra energy. But when these women from East Africa carry a load, they gradually increase their efficiency up to 75 per cent by means of this pendulum action. They reach this maximum efficiency when they are carrying about 20 per cent of their body weight. Once they carry more than 20 per cent of their body weight, they don't increase their efficiency any more, and so they begin to burn up extra calories.

Now we don't know how they do the pendulum action so well – whether it's because they use their heads, or they practise from an early age, or they have good posture, or land with a stiff out-stretched leg, or something else – but many people would like to know the secret of this high-efficiency walking. The military are very interested in having soldiers burn up less energy in a forced march – the troops wouldn't have to be fed so much, and they would arrive at their destination less tired. Many long-distance athletes, not just those who enter the walking races, are also very interested in the secret.

Scientists plan to investigate this super-efficient walking. They want to place accelerometers on various parts of these women, and combine them with high-speed video analysis, as well as force plates.

But in the meantime, you can use this knowledge to save money on food bills. Throw away your backpacks and briefcases, strap your stuff to your skull, and swing through your day like a pendulum.

References

Nature, Vol. 319, 20 February 1986, 'Making headway in Africa' by R. McNeill Alexander, pp.623–624.

Nature, Vol. 319, 20 February 1986, 'Energetic cost of carrying loads: have African women discovered an economic way?' by G.M. O. Maloiy, N.C. Heglund, L.M. Prager, G.A. Cavagna and C.R. Taylor, pp.668–669.

The Lancet, 5 December 1987, 'Fatness and the energy cost of carrying loads in African women' by Jones et al., pp.1331–1332.

Nature, Vol. 375, 4 May 1995, 'Freeloading women' by Richard Taylor, p.17.

Nature, Vol. 375, 4 May 1995, 'Energy-saving gait mechanics with head-supported loads' by N.C. Heglund, P.A. Willems, M. Penta and G.A. Cavagna, pp.52–54.

APPLE STACKING

APPLES ARE THE MOST widely cultivated fruit on our planet. You can either eat them, cook them or turn them into cider. Apples are famous in our history – a golden apple is supposed to have caused the Trojan War, and William Tell is famous for shooting the apple on his son's head.

Apples belong to the Rose family. There are about 7500 different species of apple, but only about a dozen or so appear in our shops. All modern apples seem to have evolved from the wild crab apple of Europe and Asia, which was a small fruit with a rather sour taste. Apples grow best in a temperate climate in which the temperature drops down to near freezing for a few months each year. Apple trees usually have alternate years of heavy and light crops, which are referred to as 'on' or 'off'.

An apple is a wonderful thing. Each apple has about 4 milligrams of vitamin C, about 2 grams of dietary fibre, but only 45 calories. The average eating apple weighs about 100 grams, but according to the *Guinness Book of Records*, the World Record is 1.36 kilograms for an apple grown in the United Kingdom in 1963.

Apples can be eaten raw, or cooked, baked, boiled or preserved. The more acid there is in an apple, the more easily it breaks down into a puree. You can even turn apples into a delicious drink, cider, which can be non-alcoholic or alcoholic. Unfermented alcoholic cider is around 2–8 per cent

Apples in Hospitals

In the United Kingdom, apples are not allowed to be grown in prisons or mental hospitals, because of the risk that the inmates might be poisoned by the cyanide in the the apple seeds!

Newton and That Apple

In 1666, Isaac Newton supposedly saw an apple fall from an apple tree, and so was inspired to develop his Universal Theory of Gravitation. This tree, the Isaac Newton Apple Tree, lived for some 150 years, and descendants of this tree are still grown at his home in Woolsthoorpe Manor, and in the National Physical Laboratory in Teddington.

Unfortunately for the legend, recent historical research suggests that there was no apple involved at all, and that this story was just a total fabrication!

alcohol (about the same as beer), but if you let it ferment into apple brandy it can reach 50 per cent alcohol.

The apple tree was much loved by cabinetmakers, because the wood survived well, it was very hard and it had a very fine grain. Unfortunately, these days there are problems with getting a reliable supply of this timber.

Apples were widely cultivated by the ancient Greeks and Romans, and one rather special apple supposedly caused the Trojan War!

The legend says that Eris, the Goddess of Discord, had not been invited to a royal wedding between a fellow god and a goddesses. So she sneaked into the banquet hall and threw onto the floor a golden apple on which was a message that said it was to be given to the most beautiful goddess. Three goddesses, Hera, Aethena and Aphrodite, each claimed that she was the most beautiful goddess, and so she should be given this valuable and beautiful golden apple. Zeus, the King of the gods, was the obvious god to judge which of the three goddesses was the most beautiful, but he refused to make a judgment. So the three goddesses asked Paris, the Prince of Troy, to make the decision.

Each of the three goddesses desperately wanted to win, so they each offered Paris a bribe. Hera promised that she would make him a very powerful ruler, while Athena said that she would make him famous as a brilliant military man. But Aphrodite won him over with her bribe – that he would be able to have the most beautiful woman in the world. So there and then at the royal wedding, Paris chose Aphrodite as the most beautiful goddess. For his reward, he chose a woman called Helen.

Unfortunately, Helen was already married to Menelaus, who was king of one of the Greek city-states called Sparta. So Paris sailed to Greece, where he was received as an honoured guest by Helen and Menelaus. As Aphrodite had promised, Helen soon fell in love with Paris, and eloped with him to Troy. Menelaus then raised an army with his fellow Grecian kings, and sailed off to Troy thus beginning the

Trojan War. And that's how an apple caused the Trojan War!

An apple was also a major part of the story of William Tell. In 1307, Burglen, where William Tell lived in Switzerland, was under the control of the hated Austrians. According to the legend, he refused to obey a direct order of one of the Austrian leaders. As his punishment, he had to shoot an arrow through an apple placed on his son's head. Of course, he shot the apple, not his son, and went on to lead a successful rebellion. Today, William Tell is a symbol of resistance to cruel oppression. In 1804, the German author Schiller wrote a play called *William Tell*, which Rossini later made into an opera.

Apples were colonised early in Australia's history. In 1788, the first year of European settlement in Australia, Governor Phillip planted some apple trees in Port Jackson. He had carried them safely on the First Fleet, all the way from Rio de Janero and the Cape of Good Hope. In that same year William Bligh also planted some apple trees upon landing at Adventure Bay in Tasmania.

In Australia, the peak production of apples occurred around 1969, when some 10 million trees were grown on some 400 square kilometres. But by the mid-1980s, this figure had dropped to some 6 million trees, which delivered about 300 000 tonnes of apples. This was just a drop in the bucket when compared to the worldwide production of apples, of some 41 million tonnes.

Apples feature strongly in both the spoken and written word.

'The apple of one's eye' is a person who is treasured or highly regarded. 'She'll be apples' means that things will turn out just fine. 'Apple pie' is a term used to refer to values that are especially American. To 'apple-polish' means to

Apple Dessert

The famous dessert, Apple Charlotte, was created by the French chef Marie Antoine Careme. He was called 'the King Of Cooks, and the Cook Of Kings' – he was chef to Czar Alexander I, Charles Talleyrand and George IV. When George IV was still the English Prince Regent, Careme worked for him from 1816 to 1890. His new dessert, layers of sugar and apple that were baked between slices of previously buttered bread, was called Apple Charlotte, after the Prince's daughter.

Large Apples

It seems as though large apples are a fairly new development in the history of apples. Today, when birds eat apples, they eat only the flesh, and leave the seed behind. So birds do not carry apple seeds from one place to another. Maybe back in the old days apples were so small that the birds were able to swallow them whole, and so spread the seeds around the countryside.

Granny Smith Apples

The Granny Smith apple is one of the first modern Australian foods. It's a wonderful apple – it's very good to eat, it's one of the best cooking apples, and it keeps better in storage than practically any other apple. In the early days they found that apples wrapped in oiled tissue paper could be successfully kept in cold storage for eight months.

Granny Smith was actually born Maria Ann Sherwood, in 1800, in Sussex. She arrived with her husband in New South Wales in 1837. Together with their five children, they farmed land in Ryde. By the time she was in her sixties, practically everybody called her 'Granny' because she had grandchildren, and because of her appearance – she habitually wore a quaint bonnet and an apron, and always carried an old-fashioned, two-lidded basket. Because her husband was handicapped, it was Granny Smith who always took the produce to and from the markets. There are two quite different stories about the origin of the apple that bears her name.

The first story goes like this. In the mid-1860s she brought home some cases of apples from the Sydney fruit markets. A few of the apples at the bottom of the cases were rotten, so she threw them out on the bank of a creek which ran near her house. One of these rotting apples took root and grew into an apple tree, probably around 1868. A small 12-year-old boy tasted one of the apples, and even though it was green, he said that it tasted better than any apple he had ever tried. Anyway, that was the story that he gave in 1924.

The second story was told by her grandson, Benjamin Spurway, in 1956. He says that Granny Smith was given some apples by a fruit agent, to see how well they would cook. She used them to make apple pies, and threw the apple cores, complete with seeds, out of her kitchen window. Soon, the seedling that would be the mother of all Granny Smith apples was growing next to the kitchen wall.

Either way, Granny Smith had the original Granny Smith apple tree, and she soon set up a whole orchard of them. That orchard was in what is now the Sydney suburb of Eastwood. She took the fruit to market in cases which were marked 'From Granny Smith's'.

The strange new green apple was an instant success. It has earnt more money for Australian apple growers than all other types of apple combined. In fact, in 1975, 40 per cent of all apples grown in Australia, and 50 per cent of all apples exported from Australia, were Granny Smiths.

Unfortunately, that original tree was totally destroyed. Granny Smith died on 9 March 1870, aged 70 years. She has two memorials – a tombstone in St Anne's Church at Ryde, and millions of Granny Smith apple trees all over the world.

seek favour with or seek the approval, while we all know what it means to 'bring an apple for the teacher'.

The Irish proverb 'When the apple is ripe, it will fall' refers to maturity. In the Bible, the Song of Solomon praises apples for their therapeutic value when it says: 'Stay me with flagons, comfort me with apples: for I am sick of love'. Ralph Waldo Emmerson compared apples to belief: 'We are born believing. A man bears beliefs as a tree bears apples.' And William Shakespeare wasn't very flattering to politicians or elections when he said, 'There's small choice in rotten apples'.

So why if apples are such a core part of our society have we been stacking them wrongly all this time? When apples are stacked in boxes, they're usually stacked with the cores lying horizontal, or parallel to the ground. Apple growers believed that if the apples were stacked with the cores vertical, the stalk of one apple would pierce the apple above it. But Ali A. Khan and Julian Vincent from the Biomechanics Group in the Department of Pure and Applied Zoology at the University of Reading in the United Kingdom disagree. They've done the research, and they reckon that apples stacked with the cores vertical don't bruise easily!

They set up an experimental apple-squashing device that had two heavy plates parallel to each other. The apple was placed between the steel plates and then the top plate was moved slowly down to squash the apple, at the rate of 2 millimetres per minute.

They found that if they laid the apple on its side, even a small amount of squashing left permanent bruises in the apple. But if they squashed the apple while the core was vertical, there was a fairly large range in which the apple could behave elastically. In other words, if the apple was squashed down a bit, and then the pressure was taken off, it could recover its original shape with absolutely no damage. Of course, if you gave it a really big squash and went outside this elastic range, you would cause bruising.

When apples are stored in a box, the apples on the top apply their weight to the apples on the bottom. According to

Apples in Retirement

One problem with apples is getting hold of the ones at the top of the tree – there's always the danger of falling off the ladder and breaking something important. But W.D. Campbell, who writes the Country Life column for the English newspaper the *Guardian,* planned well for his retirement. He grafted apple tree stocks onto dwarf trees, so his favourite apples grew – but on smallish, easily accessible trees.

Johnny Appleseed

There really was a Johnny Appleseed, and his real name was John Chapman. He lived from 1774 to 1847. Some time around 1801 he began his travels, and over the next 40 or so years he travelled some 160 000 kilometres, criss-crossing Pennsylvania, Ohio and Illinois. Wherever he went, he planted apple seeds.

our zoological biomechanics, the bottom apples suffer what is called 'static compression'. When the box is jolted during transportation, these apples on the bottom receive an extra load which is called 'dynamic compression'.

Imagine that you have a tennis ball between our two steel plates. As you squash it on its north and south poles, it will spread a bit at its equator. Now replace the tennis ball with an apple that is lined up with the core standing **vertically**. Imagine that you squash the apple gently at the top and the bottom, its north and south poles. It will shrink a little bit between its north and south poles, along the line of the core. At the same time, it will expand a little bit around the equator and get wider. If you don't squash the apple too much, once you lift the plate, the apple will return to original shape.

To be technical, the apple can be squashed with an energy of 0.5 joules, and recover fully, with no bruising at all – if it has been stacked with the core vertically. In fact, it turns out that your average apple, at the bottom of a case, undergoing average transportation, will never feel any energies greater than 0.5 joules. This means that it can make the trip from the orchard to your fruit shop to your mouth without any bruising at all – if it's stored vertically.

But suppose you turn the apple on its side, with the core running **horizontally**, or parallel to the ground. As you squash the apple on its equator, it tries to spread wider along the north and south poles of its core. But the core is a very stiff and inextensible bunch of fibres, so it can't expand. Result – there is no elastic range, and the apple immediately begins to bruise. In the real world of fruitology, a lot of apples arrive at your house already bruised.

So all we need is a machine to pull out the stalks, and then the apple-growers can send the apples to market with their

Non-bruising Apples

The flesh of the apple is made up of columns of cells all radiating out from the core. These cells are about 150 to 300 microns in diameter (about 2 to 4 times the thickness of a human hair).

Apples discolour because of an enzyme called PPO (polyphenolic oxidase). When an apple is bruised, the cells split open. The PPO then comes into contact with phenolic compounds inside the cells, and makes melanin pigments. Melanin gives us humans a brown skin (which we call a suntan), and it gives a brown colour to the bruised part of the apple.

CSIRO scientists have come up with a way to modify potatoes so that they make a PPO which is only 10 per cent as active as normal PPO. They intend to use this technology to make potatoes that will be spot-free – and if it works, maybe they'll try to make bruise-free apples!

cores vertical, and the unbruised fruit will be the apple of our eye.

References

The Australian Encyclopaedia, Australian Geographic Society Pty Ltd, 1988, pp.98, 100, 169–170, 390, 905, 1593, 2524, 2848–2849.

New Scientist, No. 1753, 26 January 1991, 'Apple stackers go to the core of the problem' by Simon Hadlington, p.15.

Journal of Texture Studies, Vol. 22, 7 March 1991, 'Bruising and splitting of apple fruit under uni-axial compression and the role of skin in preventing damage' by Ali A. Khan and Julian V. Vincent, pp.251–263.

New Scientist, No. 1905, 15 January 1994, 'A bite out of the fruit of knowledge' by Jack Harris, pp.47–48.

New Scientist, No. 1961, 21 January 1994, 'Wave goodbye to discoloured fruit' by Tim Thwaites, p.24.

VAMPIRES SUCK

TIME HAS FINALLY caught up with Count Dracula. His castle is crumbling and is in urgent need of repairs. Dracula was supposedly a vampire – a reanimated or 'undead' corpse. Traditionally, vampires come out only at night, and drink blood from their victims, who themselves are then turned into vampires. We all know that vampires can be slowed down with garlic, or daylight, or brought to a complete halt with a wooden stake through the heart.

Some scientists might have solved the mystery of the vampires. They traced the symptoms that vampires have traditionally displayed to a disease called 'porphyria'.

The first recorded use of the word 'vampire' in the English language is in a publication written some time before 1734. This word entered the English language from the French language (*vampire*) and the German language (*vampir*). But it originally came from the Old West Slavic word *vùmpir*.

Vampires have long been a part of Slavic folklore and mythology, but the Dracula we all know and love was based on a rather bad-tempered prince called Vlad the Impaler (also called Vlad III). He lived from 1431 to 1477, and was prince of Walachia (also spelt Wallachia and Wallacia), in what is now called Southern Romania. They called him the 'Impaler' because he would kill people by impaling, or skewering, them on wooden stakes. Altogether, he killed some 100 000 people, both enemies of his state and his own subjects, before he was killed by the Turks. His father, who was just as cruel, was called 'Vlad Dracul', meaning 'Vlad, the Devil' – so his son,

Vlad the Impaler, was called 'Vlad Dracula', meaning 'Vlad, Son of the Devil'. (Dracula also means Son of the Dragon.)

The legend of Vlad Dracula, or Vlad the Impaler, did not spread very widely. The locals knew of Vlad, because they had experienced him. Vlad was also known to Germans and Hungarians because of 15th century pamphlets written about him in those languages. Apart from that, not much was known about Vlad to the outside world.

In the 1890s, Bram Stoker, an unsuccessful Victorian writer, became aware of Vlad Dracula when he was doing research in the British Museum on his next novel. He had written many horror stories, but they had not sold well, and he made his living by managing Sir Henry Irving, the great Shakespearean actor (who was also the first English actor to be knighted). Bram was impressed by the wickedness of Vlad, and how he supposedly drank the blood of his victims. Vlad's castle still bore his name, Castle Dracula, and this suggested the obvious title for the book.

But what about the location? A Hungarian friend, a professor at the University of Budapest, suggested Transylvania, a kingdom immediately to the north of Walachia, as a location. Bram visited Transylvania, and realised that its lingering morning fogs, dark and sinister mountains, and Gothic-looking castles drenched in a history of blood made it the perfect location. He had a sure-fire winner in the legend of Dracula, and when it was published in 1897, his Gothic horror romance, entitled simply *Dracula*, was fabulously successful.

Bram Stoker set the story in both London and Transylvania. In the story, Dracula drinks the blood of beautiful young women, who then turn into vampires themselves. Eventually, a Dutch metaphysician and scientist called Van Helsing kills Dracula by impaling him with a wooden stake through the heart. It's such a good story that dozens of plays, musicals and movies have been based on it.

But how can the vampire myth be related to a disease? It can, if the disease is a disease of the blood.

A Brief History of Vlad the Impaler

In 1444, Murad II, the Ottoman sultan, forced Dracul, the Prince of Wallacia, to leave his sons Radu and Dracula (then aged 9 and 12) with him, as hostages. Murad II became infatuated with Radu, and kept him in luxurious surroundings. Dracula was not as popular, and stayed in a cellar.

In 1447, Dracula's father (Dracul) and brother (Mircea) were betrayed by their own noblemen and murdered near the capital, Turgoviste. A Hungarian leader, Janos Hunyadi, took over the throne, but was soon defeated in a battle against the Turks. This left the throne open to Dracula. He escaped from Murad II in 1448, and ascended to the throne of Wallacia. But after a few months, Vladislav II took over the throne, and Dracula spent the next eight years in exile.

He travelled to Turkey and then to Moldavia, where he stayed with Prince Bogdan II (his step uncle) until Bogdan was assassinated in 1451. He then stayed in Hungary with his old rival, Janos Hunyadi, until Janos died of the plague in 1456.

At the age of 25, he invaded and took over the throne of Wallacia. He then went on a rampage of killing against his own people, and became known as 'Vlad Tepes' – Vlad the Impaler. He ordered that 'loose' women be skinned alive, and that beggars be summoned to a 'banquet' – and then be burnt alive in the banquet hall. That same year, he invaded Transylvania, and impaled his enemies on stakes. In 1459, he avenged his father's death by inviting 500 of his noblemen to a banquet – whereupon he impaled them upon stakes.

He invaded the city of Brasov in Transylvania and impaled the entire population on stakes on a hill near the church of St Bartholomew – and then ate his meal surrounded by a forest of his dead and dying victims! He impaled 10 000 citizens of the city of Sibiu, and 20 000 citizens of Fagaras and Amias. When ambassadors of the Turkish sultan would not remove their turbans in his presence, he had the turbans nailed to their heads! In 1462, he killed 20 000 Turks along the River Danube, but lost his throne – which was taken over by his pro-Turkish brother, Radu (supported by the new sultan, Mohammed II). Vlad was kept as a prisoner for four years by King Mathias of Hungary. According to legend, his prison guards gave him small animals (birds, mice etc), which he tortured to death, and then impaled.

In 1476, he again ascended to the throne of Wallacia. But the very next year, he was ambushed outside Bucharest and beheaded by the Turks. In modern-day Romania, he is revered – because he protected the nation against the Turkish invaders. The story told by Bram Stoker is considered insulting to the memory of Vlad, who fought hard to protect the country's independence.

We humans all have blood, and one of the important jobs of the blood is to carry oxygen from the lungs, and unload it wherever it's needed. The molecule in the blood that actually carries the oxygen is haemoglobin, which is carried inside the red blood cells.

It's a fairly complicated process to make haemogloblin. Our bodies start with the amino acid called 'glycine', and then modify and change it in some eight successive steps. Some of the chemicals in the intermediate steps are called 'porphyrins'. These porphyrins are hollow molecules. In the last step of this process, an atom of iron is inserted inside the hollow molecule, to make 'haem'. Finally, 'globins' are then added to make 'haemoglobin'.

The eight-step process to make the 'haem' molecule is a bit like a factory production line. If there is a hold-up or blockage at any stage, products from the earlier stages will begin to pile up and accumulate. A disease which causes a build up of porphyrins such as this is called a 'porphyria'. There's a whole family of diseases called porphyrias in which the sufferer cannot make enough haemoglobin.

The theory is that in Slavic Europe, some people suffered from one of these porphyrias (probably a type called Variegate Porphyria), and so started the ancient legend of vampires.

People suffering from porphyria are not very good at making haemoglobin. So where would they have got blood from before transfusions were invented? Maybe they would have drunk the blood of animals...or even human blood.

Now what about garlic? Everyone knows that vampires don't like garlic, and the theory might be able to explain this as well. Red blood cells last about 120 days before they are broken down by an enzyme called P450. It turns out that garlic stimulates this enzyme, so old red blood cells die sooner. It is said that garlic could easily be fatal to a poor, misunderstood porphyria-sufferer who had *just* been getting by on the marginal amount of haemoglobin that they had.

What about vampires wandering about only at night?

British Royal Family had Porphyria?!

It's still a little controversial as to whether various members of various British Royal Families really suffered from one of the porphyrias.

In 1969, Ida Macalpine and Richard Hunter, in their book *George III and the Mad-Business* strongly suggested that George III probably had Variegate Porphyria. They examined the long and complicated family tree very carefully. They unearthed much descriptive evidence about the illnesses of James I and his cousin Arabella Stuart, which seemed consistent with their having had porphyria. They also collected evidence that many descendants of James I may have been affected. They even examined the stools and urine of four living descendants to confirm the diagnosis of porphyria. However, because these four people were 'of royal descent', they were unable to supply names!

Geoffrey Dean, a world expert on porphyria, disagreed! He has built his reputation by following porphyria in closely related families in South Africa. In fact, he is sometimes called 'Mr Porphyria'! In a book published in 1973, he denied that any of the signs and symptoms described by Macalpine and Hunter are due to a single disease such as porphyria. He suggests that perhaps George III had cerebral syphilis, Mary Queen of Scots had gastroenteritis, James I had a kidney stone, and so on.

However, there is one piece of evidence that suggests that Geoffrey Dean may be wrong, and that Macalpine and Hunter may be right.

Dr Johannes George von Zimmerman, a Swiss doctor, treated many sufferers of porphyria, such as Caroline Matilda, Augustus, and Frederick the Great. A doctor did not get to such an exalted position by being stupid, or unobservant. He wrote to George III, saying: 'It has come to our notice that other members of the Royal Family, including His Royal Highness the Duke of York and Prince Edward (Duke of Kent), are subject to the same paroxysms, and this arouses our suspicions of a hereditary predisposition.'

Dr Zimmerman certainly thought that these symptoms which afflicted many members of the Royal Houses of Europe, were connected — and he was a very careful observer.

Sufferers from Variegate Porphyria have an excess of the porphyrin molecule all over the body, including *in* the skin. It turns out that light excites this molecule, which then attacks the cells of the skin, causing severe scarring and blistering – and therefore the sufferers of this disease don't like to go out in the sun.

This porphyria can have a few nasty side effects. The sufferers get red fluorescent teeth (which would give the impression that they had recently enjoyed a meal of blood!). The body becomes very hairy, even on the forehead. The build-up of chemicals interferes with some of the normal workings of the nervous system – and sometimes the normal movements of the gut stops. This gives them severe tummy pains, which makes them pretty mean. They can become paranoid, or frankly even insane. Porphyria can be caused by drugs (such as barbiturates) or toxins (such as lead), or it can be inherited. If it has been inherited in a mild form, taking drugs or toxins can set it off.

There is a connection between this vampire disease and both Vlad the Impaler and the British Royal Family (both of whom have castles in need of urgent repairs).

It seems (according to some scientists) that the British Royal family, as well as various European Royal families, has had this vampire disease (or at least one of porphyrias) running through them for centuries. Clinical evidence consistent with porphyria, and evidence of discoloured purplish urine, caused by an excess of porphyrins, has been noted in James VI, George III, Frederick the Great and Augustus, Duke of Sussex. Clinical evidence of porphyria has been noted in Mary Queen of Scots, Queen Anne, Frederick William I and George IV, as well as Caroline Matilda, Queen of Denmark.

James I supposedly was proud of the fact that whenever he drank port wine, it would come out the same colour as it went in (because of the porphyrin molecules in it).

George III had five major acute attacks of porphyria – an attack from January to July in 1765, one from October 1788

Vampire Bats

Now there are a lot of animals that don't have a very nice reputation, and one of these is the vampire bat! Vampire bats sneak up on you at night, use the heat-seeking cells in their nose to home in on your blood vessels, slice open your skin with their razor-sharp teeth, and spit into the open wound! Their saliva contains a chemical that keeps the blood flowing freely, and over the next 25 or so minutes, they lap up your precious red liquid of life. These greedy little critters will drink so much blood at one sitting (even as much as their own body weight) that sometimes they can't even fly away, and have to sit around for a few hours to digest the meal before they can take off.

There are three types of vampire bats that drink blood — the white-winged vampire bat, the hairy-legged vampire bat, both of which mostly feed on the blood of *birds,* and the common vampire bat, which feeds mainly on the blood of *mammals.*

But there's always been a mystery about these blood-thirsty little mammals. If they're so *greedy,* how come they're so *kind* to each other? You see, every now and then, a well-fed vampire bat will give a hungry vampire bat a feed by vomiting up some of its recently stolen blood from its stomach.

As Charles Darwin noted in his book *Descent of Man,* co-operation was important in the evolution of the human race. He wrote: '...his want of natural weapons and so on, are more than counterbalanced by his...social qualities which lead him to give and receive aid from his fellow-men.' But for vampire bats, this mutual support has a far more immediate impact.

Feeding is crucial for all living creatures. In the case of the vampire bats, if they go 60 hours without a feed, they'll starve to death!

In one study, about 30 per cent of the bats under two years of age were unable to get a nice drink of blood on their average night out. Of course, as they got older they got better at it, and mature bats had only a 7 per cent failure rate in getting a drink. If you do a *computer simulation* based on these numbers, it works out that in any given year, about 82 per cent of the vampire bats *should* die. But when you look at a population of real vampire bats in the wild, only 24 per cent actually die each year!

It all hinges on the bat's buddy system.

In one five year study, Gerald S. Wilkinson, an assistant Professor of Zoology at the University of Maryland at College Park, looked at this cute habit of vampire bats giving each other a drink, by vomiting up blood for each other. There seemed to be two major categories of blood regurgitation. Either the bats involved were very closely related to each other, or else they had matched up with a buddy, so that they would vomit up blood *only* for each other. ➤

▷ Now the individual vampire bat doesn't really care that if he or she helps out another vampire bat, their group will have only a 24 per cent death rate, rather than an 82 per cent death rate, each year. What the individual bat that gives up some of its blood to another bat is worried about is whether it will be so weakened that its own life will be at risk.

But Professor Wilkinson, after doing a cost-benefit analysis of blood sharing, found that the gain was greater than the loss incurred. Take the case where a vampire bat vomits up 5 millilitres of blood for its hungry buddy. Well, the donor bat would, on average, lose only *six* hours out of the *60* hours that it had left until it starved to death. But the bat that drank the blood gained 18 hours of extra life, and so it benefited much more than the donor bat lost!

I guess that in this case, if you are generous to your friends, you are not just a sucker being taken for a ride!

Castle Dracula

Castle Dracula was built around 1370, at Bran, in the Carpathian Mountains, in what is now Central Romania. The foundations and the two lower floors were carved out of the rock of the mountain. The castle attracts some 4000 tourists each day.

By late 1994 there were large cracks, several metres long, in the cliff on which it stands. Some $US 300 000 is needed to repair the castle. For a long time it was considered to be tarnishing Vlad's memory to organise a 'Dracula stage show'. However, as the repairs became more urgent, and after five years of capitalism in Romania, the world's first Dracula Congress was held in May 1995. This government-sponsored congress lasted five days. It culminated in an eerie masked ball on the slopes of the mountain. Money should soon be available for the repairs.

to February 1789, another from February to March 1801, a brief attack from January to March 1804, and a long attack from October 1810 until his death on 29 January 1820. He also had many minor attacks over the years. The first acute attack lasted about six months and was so disabling that Parliament passed a Regency Bill, which would allow his son to take over. He recovered when the Bill was with the House of Lords. However, he never really recovered from his last attack of porphyria in 1810.

FAIRLY INTERESTING MOMENTS IN VAMPIROLOGY #23

Royal Flush

PALACE TOILET DETAIL TAKING CARE OF GEORGE THE TURD.

BEATTY 95

The porphyrin molecules are found in the faeces, because the body is trying to get rid of them. When these molecules in the faeces are hit by ultraviolet light, which is found in sunlight, they fluoresce, but only for an instant. This is what King George III might have called a Royal Flash in the Pan!

References

This story is dedicated to Dr Vivian Whittaker, from the University of Sydney, who told me and some 250 other first year medical students this wonderful story.

British Medical Journal, 8 January 1966, 'The "Insanity" of King George III: a classic case of porphyria', by Ida Macalpine and Richard Hunter, pp.65-71.

British Medical Journal, 13 January 1968, 'Historical implications of porphyria' by John Brooke, pp.109–111.

Medical History, Vol. 26, 1982, 'Porphyria revisited' by Lindsay C. Hurst, Vol. 26, pp.179–182.

New Scientist, No. 1329, 28 October 1982, 'The curse of Dracula', by Lionel Milgrom, p.244.

New Scientist, No. 1332, 18 November 1982, 'Old werewolves' by L.S. Illis, p.459.

New Scientist, No. 1407, 26 April 1984, 'Vampires, plants, and crazy kings' by Lionel Milgrom, pp.9–13.

Extraordinary Origins of Everyday Things, Harper and Rowe, New York, 1987, 'Dracula' by Charles Panati, pp.179–180.

Medical Observer, 30 September 1994, 'Did a mutant enzyme make George III mad?' by Chris Smith, pp.60, 61.

Scientific American, February 1990, 'Food sharing in vampire bats' by Gerald S. Wilkinson, pp.76–82.

NUCLEAR COAL

What is Coal?

IT WAS THE CHINESE who first used coal as a fuel back in 1100 BC. Even today, most of the electricity in the world is made by burning coal. In 1990, 66 per cent of the world's electricity was made by burning fossil fuels, while nuclear power accounted for only about 15 per cent. But according to Alex Gabbard of the Metals and Ceramics Division of the Oak Ridge National Laboratory in the USA, a coal-fired power plant gives off more radiation than a nuclear-fired power plant!

The process of making electricity is almost identical in a coal-fired power plant and a nuclear power plant. In each type of plant, heat turns water into steam. The steam is blasted onto the blades of turbines, which then spin. The turbines are attached to generators or alternators, which make the electricity. (There is one minor exception to this rule that steam turns the turbine blades. A helium-cooled nuclear reactor uses helium gas, not water/steam, to turn the turbines. In fact, a helium-cooled reactor is about twice as efficient as a water/steam cycle nuclear reactor, and is thus cheaper to run.)

In a coal-fired power plant, the heat to turn water into steam comes from burning coal. In a nuclear power plant, the heat comes from the splitting of atoms of uranium, thorium or plutonium. This splitting process, called fission, happens when enough fissionable nuclear fuel is placed close enough together, so that a nuclear reaction happens. The difference between a coal-fired power plant and a nuclear-fired power

According to the encyclopediae, coal is a rock which was originally plant or vegetable matter. It was mostly laid down between 140 and 390 million years ago, and then decayed in the absence of oxygen. After being compressed underground as the result of great pressure and temperature, it turned into coal.

There are many steps in the evolution of coal, and there are many ways to classify coal (age, location etc). One such way measures the carbon content. Peat has the lowest carbon content, followed by lignite (or brown coal, of which there are vast beds in Victoria). Bituminous coal has a higher carbon content. The highest of all is anthracite – about 92–98 per cent carbon.

Burning Coal Gives You Everything

Coal chemistry is a vast cauldron of witch's brew. One obvious result is vast amounts of the greenhouse gas carbon dioxide being released into the atmosphere. There are many carcinogenic chemicals made when coal is burnt. Oxides of sulphur are known to cause acid rain, and oxides of nitrogen cause breathing problems, and affect ozone. Look at arsenic, which is present in coal at about 5 parts per million. That works out to 26 000 tonnes of arsenic released into the biosphere each year.

Uranium-238 released from coal by burning can go up a chimney as uranium-238, but can land as plutonium-239. Our atmosphere is loaded with neutrons, which come from cosmic rays colliding with our atmosphere. When a neutron hits uranium-238, it converts it into plutonium-239, which is more toxic than uranium. However, the amount ➤

plant is what provides the heat. But what's the same is that in each type of electrical power plant, heat makes steam, which makes electricity.

Now coal is a very impure fuel. It's mostly carbon, but there are impurities like silicon, aluminium, calcium, magnesium, sodium, titanium, arsenic, potassium, sulphur and mercury and tiny amounts of uranium and thorium. In fact, 73 different elements have been identified in coal! On average, coal has 1.3 parts per million of uranium and 3.2 parts per million of thorium. (Until modern analytical methods were invented, these tiny amounts were simply too small to be measured.) These are very small quantities, but on the other hand, a lot of coal gets burnt.

To run your average 1000 megawatt coal-fired power plant for one year, you need to burn about 4 million tonnes of coal. That 4 million tonnes of coal contains 5.2 tonnes of uranium and 12.8 tonnes of radioactive thorium – as well as 0.22 tonnes of radioactive potassium-40. But that's just from a single 1000 megawatt plant in just one year. The worldwide use of coal in 1991 was about 5100 million tonnes. When that coal was burnt, some 6630 tonnes of uranium and 16 320 tonnes of thorium were released into the biosphere.

That 6630 tonnes of uranium included over 47 tonnes of uranium-235 – the stuff that goes bang. That 47 tonnes of uranium-235 could be made into some 1700 World War II-style atom bombs, with a total combined explosive yield of 34 megatonnes. In fact, just a single 1000 megawatt coal-fired power station releases enough uranium-235 to make a World War II-style atom bomb each year. (The yield of nuclear weapons is measured in tonnes of TNT. As an aside, one tonne of coal has three times the energy of a tonne of TNT. But the TNT is better as an explosive, because it burns much faster.)

If you look at the amount of coal that is predicted to be burnt in the 100-year period from the year 1937 to the year 2037, you're looking at 640 billion tonnes of coal. That enormous pile of coal contains about 830 000 tonnes of uranium, 2 000 000 tonnes of thorium and 35 000 tonnes of potassium-40 – all of it free to enter the biosphere! We are still not too sure where it all goes. Uranium and thorium are not very mobile, but potassium-40 can easily enter the food chain.

If in the year 2037 there are 8 billion people on the planet, that works out that for every person there will be a paddle pop's worth of uranium (about 100 grams of uranium) and three paddle pop's worth of thorium – all thanks to coal-burning.

When coal is burnt, the carbon combines with oxygen. The products of this reaction are heat (used to make electricity) and carbon dioxide. But in coal there are also various inorganic impurities which don't get burnt. These impurities turn into coal ash, which can make up between 3–30 per cent of the weight of the coal, depending on the type of coal.

There are two types of coal ash – bottom ash and fly ash. In 1975, the USA burnt 410 million tonnes of coal – leaving 63 million tonnes of ash, which was made up of 41 million tonnes of fly ash and 22 million tonnes of bottom ash. That's enough ash to cover an area 12 kilometres by 12 kilometres to a depth of 30 centimetres – and that was only one year's worth!

The bottom ash tends to be made from coarser, heavier particles, which don't get lifted up into the chimney. It's called bottom ash because it's collected from the bottom of the boiler.

Fly ash is the stuff that 'flies' up the chimney. It can make up between 10–85 per cent of the coal ash – depending on the type of burner, the type of boiler, the type of coal, etc. This ash is about 50–90 per cent glass. The glass forms as silicon melts during the burning of the coal. The glass is mostly in the shape of tiny balls, between 0.5–100 microns in diameter (for comparison, a human hair is about 70 microns in diameter). It turns out that heavy metals like uranium tend to stick to these microscopic balls of glass that make up most of the fly ash. This ash is actually richer in uranium and thorium than the original coal, because while the carbon content was reduced during the burning, the amounts of uranium and thorium stayed the same, and so make up a greater portion of

> **of plutonium added to the biosphere via this pathway is probably insignificant.**

Uses for Coal Ash

At the moment, coal ash is being used mainly as landfill, in concrete and cement, in roads and pavements, and in bricks. There have been a few specialised low-volume uses. Cenospheres have been used to make a tape for fire-proofing and insulating high-voltage cables, and also to make a closed-pore insulation material for the space shuttle. Fly ash has also been used to improve the yield of rice paddies.

Where is Coal?

It's hard to get consistent figures about who has the world's recoverable coal reserves. According to the World Energy Council, China leads with 45 per cent, followed by the USA with 17 per cent, the former USSR with 12 per cent, and South Africa with 5 per cent. Australia has about 4 per cent of the world's coal reserves. It is also the largest coal exporter in the world. In 1993, Australia exported 131.8 million tonnes of coal. That coal contained about 171 tonnes of uranium. It also contained the so-far untapped vast wealth of 164 000 tonnes of titanium, 450 000 tonnes of magnesium, 2 000 000 tonnes of iron, and 3 480 000 tonnes of aluminium.

Coal and Society

Coal has long been associated with violent social change. In 1778, Scottish coal miners began to gain their freedom from the oppressive conditions which they had been forced to accept. In the USA, severe coal strikes occurred in 1884, 1897, 1897, 1902, 1914, and 1922. In many cases, coal miners were killed. On 17 July 1902, George F. Baker from Philadelphia and Reading Coal and Iron indicated publicly that he did not agree that workers had the right to look after their own interests. He said: 'The rights and interests of the laboring man will be protected and cared for – not by the labor agitators, but by the Christian men to who God in his infinite wisdom has given the control of the property interests in this country...'

the fly ash.

Nowadays, coal-fired power plants have precipitators on the chimneys to catch the fly ash, but these precipitators are only about 99.5 per cent efficient. So a small amount of fly ash, contaminated with radioactive metals, does escape. Now it's only a *very small* amount of fly ash that escapes, but your average power plant burns up a *lot* of coal.

As well as the bottom ash and the fly ash, though, there's another source of radioactivity from a coal-fired power plant. The radioactive gas radon-222 goes straight up the chimney when the coal is burnt. The precipitators that catch fine particles have no chance of capturing a gas.

When you add up all the radioactivity released from a coal-fired power plant, you find that a coal-fired power plant dumps much more radioactivity into the biosphere than a nuclear-fired power plant. According to the United States National Council on Radioactivity Protection and Measurements, the radiation exposure from an average 1000 megawatt power plant comes to 490 person-rem/year for coal-fired power plants and 4.8 person-rem/year for nuclear-fired power plants. In other words, your average coal-fired power plant puts out about 100 times more radiation than a nuclear-fired power plant!

Of course, that factor of 100 just looks at the nuclear-fired power plant by itself. It doesn't include the complete nuclear fuel cycle, which starts with ore mining, goes to fuel processing and operation of the reactor, and finishes with waste disposal. In that case, the radiation dose per citizen from a nuclear-fired power plant rockets up to 136 person-rem/year. So, according to the Oak Ridge National Laboratory figures

THIS IS A LUMP OF COAL
IT WEIGHS 500 GRAMS
IT CONTAINS .0003 GRAMS OF URANIUM
THAT'S NOT VERY MUCH, IS IT?
HAH!

THIS IS THE CUTE FURRY
LITTLE POSSUM THAT LIVED
IN YOUR CHIMNEY UNTIL YOU
FIXED UP THE FIREPLACE
AND STARTED TO BURN COAL

OW

THIS IS THE SAME CUTE
FURRY POSSUM AFTER YOU
FIXED UP THE FIREPLACE
AND STARTED TO BURN COAL

THIS IS THE HIDEOUS GIANT
MUTANT POSSUM THAT WILL
DESTROY YOUR HOUSE AND RIP
OUT YOUR LIVER BECAUSE
YOU FIXED UP YOUR FIREPLACE
AND BURNT COAL

SO **DON'T BURN COAL IN YOUR FIREPLACE**
(GET A NUCLEAR REACTOR - IT'S SAFER)

BEATTY '95

A Brief History of Coal

In 301 BC, the Greek philosopher Theophrastus wrote of rocks 'that are called coals...which kindle and burn like woodcoals'.

In 852 AD, coal is first mentioned in reference to one Wilfred, who had rented some land from the Abbot of Ceobred, and who, each year, had to send to the monastery '60 loads of wood, 12 loads of coal, 6 loads of peat'.

In the year 1306 AD, a man was executed for burning coal in London.

In the 1600s, coal became more popular as firewood became scarce and expensive.

In 1856, the first synthetic dye in the history of the human race was made by an English chemistry student, William Henry Perkin. All previous dyes had been made from vegetable, animal or mineral sources. This mauve dye was made from coal tar.

(that include the complete nuclear cycle), you'll still get over three times more radiation from a coal-fired power plant, than from a nuclear-fired power plant. A coal-fired power plant looks even less attractive when you include the carcinogenic chemicals created by the burning of coal.

Now, at the moment, nobody really thinks much about coal waste. On one hand, it might become the clean-up nightmare of the future, but on the other hand, coal waste has many potentially useful metals in it.

For example, these radioactive metals that are present in coal are loaded with energy. In fact, there is one and a half times as much energy in the radioactive metals as there is in the carbon! That potential energy is just being wasted. In fact, the uranium alone is enough to run 125 nuclear power plants for 30 years!

Just think about nuclear leakage. In 1982, some 111 nuclear-fired power plants in the USA consumed about 540 tonnes of nuclear fuel. In that same year, coal-fired power plants released over 800 tonnes of uranium (and about 1970 tonnes of thorium) into the biosphere! If a single nuclear-fired power plant released 8 kilograms of uranium into the biosphere, there would (quite rightly) be an enormous outcry.

The main problem is that in the mid-1990s, the **nuclear content of coal** has not yet reached general public awareness, in the same way that **the greenhouse effect**, **AIDS**, or **the ozone hole** have. Scientists first suggested back in 1954 that burning coal could release radioactivity into the environment. Forty years later, hardly anybody thinks of coal-fired power stations as a source of radiation – there are no nuclear regulations about the disposal of the coal ash. While the coal ash is waiting for disposal, it's not even covered to stop it from blowing in the wind, or seeping into the ground water.

At the moment, one problem is how to get the metals out of the coal ash. But there is a possible solution. We do know from our study of botany, that some plants love heavy metals so much that these heavy metals can make up to 3 per cent of their total plant weight. Maybe genetic engineering could breed for us some bacteria or plants that we could sow around coal waste, and then reap a bumper harvest.

References
Energy and Power – Scientific American, 1971, pp.3, 15, 27, 33–34, 109, 116.
Radiation Protection Dosimetry, Vol. 4 No. 1, 22 February 1983, 'The radiation dose from coal burning: a review of a pathway and data' by J.O. Corbett, pp.5–19.
Oak Ridge National Laboratory Review, Nos 3 and 4, 1993, 'Coal combustion: nuclear resource of danger' by Alex Gabbard, pp.24–32.
Microsoft ® Encarta. Copyright © 1994 Microsoft Corporation. Copyright © 1994 Funk & Wagnall's Corporation. (CD-ROM).

Weird Fly Ash

About 20 per cent of the volume (5 per cent of the weight) of fly ash is a special type of fly ash called 'cenospheres'. They are hollow balls of glass, about 20–200 microns in diameter, filled with nitrogen and carbon dioxide. They are actually less dense than water, with a relative density of 0.4–0.8 (a litre of water weighs 1 kilogram; a litre of fly ash weighs 0.4–0.8 kilograms). When the ash is tipped into storage ponds, these balls will therefore float on the surface.

LEFT VERSUS RIGHT

Famous Lefties

Left-handers include Billy the Kid, Leonardo da Vinci, Benjamin Franklin, Lewis Carroll, Pablo Picasso, Charlie Chaplin, Prince Charles, Marilyn Monroe, Bob Dylan, Ringo Starr, George Bush, Robert De Niro, Phil Collins, Paul McCartney and even the greatest guitarist of all, Jimi Hendrix.

In the 1992 United States Presidential election, all three candidates were left-handers. However, in American political history, no left-handed president has ever been re-elected, so President Clinton doesn't look like a hot favourite for 1996. (But really, it's a pretty weak use of statistics.)

MOST HUMANS ARE right-handed, but about 10 per cent of us are left-handed. There are many different mammals around, but these mammals tend to have a 50-50 split between being right-handed – or -footed or -clawed or -flippered – and being left-handed, -footed, -clawed etc. We humans are the only mammal that is not split 50-50 between lefties and righties. For thousands of years, we've been trying to work out why so many of us are right-handed. And in our long search to try and find out why we are so unevenly split between the left- and right-handed, we have also found that the sub-atomic universe is not symmetrical!

Now about 90 per cent of us are right-handed, and about 10 per cent of us are left-handed. But it's not that clear cut, because some of us who are left-handed use only the left hand, while others can use both their right and left hands. For example, a former Australian cricket captain, Allan Border, uses his left hand for batting in cricket, for holding a racquet in tennis, and for swinging a golf club; but he uses his right hand for holding a knife, and for writing and shaving. There's a lot of variation in human handedness. In general, according to a report in the *British Medical Journal* by Professor Bryan Turner, 90 per cent of people are right-handed, 80 per cent are right-footed, 70 per cent are right-eyed, while only 60 per cent are right-eared.

There seems to be a strong family link in right-handedness and left-handedness. If two right-handed people have children, there's a 90 per cent chance that the kids will be right-handed. But the children of two left-handed parents have only a 40 per cent chance of being left-handed. So it seems as though there must be other factors besides family history that control left-handedness and right-handedness.

Some people have said (and they may well have got it wrong) that left-handedness is caused by infections or shocks or trauma to the pregnant mother, while other people have tried to blame imbalances in various sex hormones in the growing foetus inside the womb. In the mid-1990s, though, we still don't have the definite answer.

In general, most of our languages are pretty cruel to left-handers. Our English word 'left' comes from the Celtic word *lyft*, which means 'weak' or 'broken'. On the other hand, the Celtic word for 'right' means 'strong' or 'straight'. The Latin word for 'right', *dextra*, has given us our English words 'dextrous' and 'dexterity'. But the Latin word for 'left', *sinistra*, has given us the word 'sinister'. In French, the word for 'right', *droit*, has given us the word for 'law', while the French word for 'left', *gauche*, gives us the word 'gawky', which means awkward or uncoordinated. In the Gypsy language, Romany, the word for 'left', *bongo*, also means 'evil'.

We talk of a favoured person sitting at the right hand of a great person, and we call some of our religious ministers the Right Reverend. On the other hand, being born on the left side of the bed used to mean being illegitimate. Left-handed compliments are supposed to be insincere compliments.

Many of our religious texts say bad things about left-handed people. Satan, who is often shown as being left-handed, supposedly sits on God's left. Those of Buddha's followers who have gained sufficient enlightenment to achieve Nirvana, travel along the right-hand path. And in the Bible, when God separated the wicked goats from the nice sheep, he sent the wicked goats to the left: 'God set the sheep on his right hand, and the goats on the left...saying unto them on the left hand "Depart from me, ye cursed, into everlasting fire" '. However, elsewhere in the Bible, a passage in Judges

Lefties and sport

Left-handers are actually at an advantage in the sport of baseball. Left-handed baseballers are called 'southpaws' because of one particular Chicago ballpark, where the pitcher's mound faced to the west. A left-handed pitcher who was facing left would have his left hand on his south side, and that is how the name 'southpaw' came into existence!

Because they're already facing the runner on first base, left-handed *pitchers* have an advantage. Left-handed *hitters* are also at an advantage, because they stand a full two steps closer to first base.

Many lefties do well in other sports, such as tennis, because of the element of surprise – their opponents are used to playing right-handers. But in some sports such as polo and jai alai, left-handers are banned entirely because they are thought to be too dangerous.

20:16 rather surprisingly says something nice about left-handed people: 'Among all this there were 700 chosen men left-handed; everyone could sling stones at an hair breadth, and not miss'.

Because of all of this prejudice that's built into our languages, as well as the generally mean comments from various religious texts, it not surprising that many cultures look down upon left-handers.

In many cultures, the right hand is for eating, while the left hand is for unpleasant and unmentionable bodily functions. In those cultures that practice amputation, wrong-doers are punished by having their right hand sliced off, thus forcing the guilty person to use their unclean hand for eating food. There have been cases where parents of a left-handed kid have had a surgeon cut through tendons in their child's left hand, so forcing them to use their right hand. There was even the famous case of the Indonesian holy man who was so

convinced that the left side of his body harboured evil that he spent his whole life learning how to breathe only through his right nostril.

Left-handers seem to suffer from a slightly different set of diseases, as compared to right-handers. They are more susceptible to some of the auto-immune diseases such as asthma, dermatitis, hay fever, rheumatoid arthritis and ulcerative colitis.

There were some studies in the late 1980s which claimed to show that left-handers had a shorter life expectancy, but they seem to have been based on bad statistics, and now nobody believes these studies. Some studies also claim to show that left-handers are more likely to have accidents than right-handers. For example, a US Navy study of some 5000 sailors showed that left-handers were 34 per cent more likely to have serious accidents. But that might not have anything to do with the innate properties of being a left-hander, but rather reflects that the world is designed for right-handers. After all, when was the last time you saw a left-handed cheque book, or left-handed scissors, or even a left-handed tin-opener?

In some cases, left-handers seem to have a slight advantage, because of the slightly different wiring of their brain. Now the first thing to realise about the human brain is that we don't have just one brain in our skull; we actually have two quite separate brains – a left brain and a right brain. In general, the left brain controls speech, while the right brain controls non-verbal skills such as drawing, and imaging shapes.

These separate brains are joined by a bundle of nerve fibres called the corpus callosum. Inside the average skull there are about 10 billion nerve or brain cells, or about 5 billion brain cells in each

Left-hand Drive

The reason that those of us who live in the United Kingdom and Australia drive on the left is because of Pope Boniface VIII. Way back in 1300 AD, he said that any pilgrims who were travelling to or from Rome should always keep to the left. Gradually this habit spread across the world, including England. England is a fairly small country, with smallish roads, and the English kept this habit as they developed their road system.

But as the Industrial Revolution began in the 1700s, people in most of Europe and America began to use road transport to haul much larger loads than were commonly moved in England. They used pairs of horses side by side to pull big wagons. The driver would sit on one of these horses, and would usually use his right hand to whip the horses along. So if he was right-handed, and he wanted to be able to whip a horse in either the left or the right row, that meant he had to sit on the left rear horse, so that his whip hand (his right hand) would be between the two rows of horses. But because he was sitting on the left rear horse, he had to move the horses and the wagon over to the right side of the road so that he could get a clear view of the road ahead from this position. Well anyway, that's one theory.

half-brain. But there are only about 200 million cells joining your left brain and your right brain. In other words, only 2 per cent of your total nerve supply is used to join your two half-brains together so that they can talk with each other. It turns out that both left-handers and ambidextrous people (those equally able to use their left and right hands) have a corpus callosum about 10 per cent bigger than your average right-hander – which might mean that they are slightly better at getting their two brains to talk to each other.

But left-handers might have another advantage over the righties when it comes to surviving a stroke. About 99 per cent of right-handers have the control centres for speech in their left brain. This means that if they have a stroke which affects the speech area in their left brain, they can't speak any more. But the situation is different for left-handers. Only about 60 per cent of left-handers have their speech centre in their left brain; 40 per cent of the left-handers have it shared between their left and right brains. That lucky 40 per cent of the lefties who have the speech centres shared between the two almost separate sides of the brain has a slight advantage when it comes to surviving a stroke. If a speech centre in one brain gets knocked out by a stroke, they've got a good chance of regaining speech by using the speech centre in the other brain.

There is yet another left-right difference in brains between people – but this time it's between men and women. In one study, the brains of men and women were scanned while they were solving word games. The men would tend to use only the left side of their brains to solve their problems, while the women would use both sides of their brains!

In general, the person who uses both sides of their brain would have greater recovery from a stroke than a person using only one side. So it seems that as far as speech and solving word games is concerned, the best survivor of a stroke might be a left-handed woman.

Now philosophers have wondered about left-handedness and right-handedness for a long time, but it was a 25-year-old scientist, Louis Pasteur, who in 1848 found left-handedness and right-handedness in science.

He was looking at crystals of tartaric acid with his microscope, and he suddenly realised that the crystals were not identical. They seemed to be mirror images of each other, like a right hand and a left hand. So, with fine tweezers, he painstakingly and tediously separated the jumbled-up mixture of crystals, so that he finished up with one bowl full of one type of crystals, and another bowl full of their mirror-image crystals. He dissolved each set of tartaric acid crystals in a separate beaker of water. When he shone a polarised light through each solution, he found that one solution rotated the polarised light to the left, while the other solution rotated it to the right. He had discovered that Mother Nature sometimes has two hands! He soon found out that many other chemicals had two mirror image forms.

In 1857, he found that a fungus had grown in one of the solutions he was working with. He knew that this solution was optically inactive – it did not rotate light to the left nor to the right – because it was an equal mixture of left-handed and right-handed chemicals. When he was confronted with this 'contaminated' solution, he decided not to throw it away. Instead, he decided to see how it twisted light.

The solution that the fungus had been living in was now optically active – it rotated polarised light! The fungus had eaten all the chemicals of one particular handedness, for food. This left the solution full of chemicals of the other form of handedness only – which was why it now rotated polarised light.

He suddenly realised that the very chemicals that make life possible on our planet, the Chemicals of Life, have a preferred handedness. And in a remarkable statement he said: 'Life as manifested to us is a function of the asymmetry of the universe, and of the consequences of this fact...The universe is asymmetrical'.

He was absolutely right that the Chemicals of Life are not symmetrical. For example, look at the chemical limonene,

Do only Humans Favour One Hand?

It is claimed that all animals will show an equal tendency to use either hand or foot, and that humans are the only animal to show a preference for one hand. However, most parrots show a distinct preference for the left foot. Common marmosets (*Callithrix jacchus jacchus*) may also show a preference for one hand. According to one study on 23 marmoset families, 43.5 per cent were right-handed, 23.9 per cent were left-handed, while 32.6 per cent were ambidextrous.

Body Symmetry

The human body is not symmetrical. There are several organs that exist only on one side – the heart, the spleen and the liver. The female breasts are usually different in size, while the right testicle is usually larger and higher.

One famous case relating to body symmetry involves a pair of twins, Leon Kane-Maguire and his brother Noel Kane-Maguire. They are both professors of chemistry – Leon at the University of Wollongong, and Noel at Furman University in the USA.

Leon is 'Noel' spelt backwards. Leon has the medical condition called *situs inversus*, where his internal organs are the mirror image to that most of us have. And his heart is on his right side.

Both the brothers work in the field of left-handed and right-handed chemistry! Leon is especially interested in identifying the different properties and uses of single-handed (left or right) drugs.

which is found in oranges and lemons. In one handedness, it smells like oranges, but in the other handedness, it smells like lemons!

A very tragic case of just how different the left- and right-handed forms of a chemical could be arose in the early 1960s with the drug thalidomide. Thalidomide, as it was sold then, contained equal amounts of both mirror-image forms. But while one form of handedness (the form tested) prevented nausea and vomiting during pregnancy, its mirror-image twin caused birth defects. Worldwide, some 10 000 babies were born with severe malformations of the limbs.

Today, we know about this difference, and so one form of handedness of thalidomide (the one that when tested did not cause birth defects) is being used to treat both leprosy and the very serious immune reaction called graft-versus-host disease, which overcomes about half the recipients of bone-marrow transplants. Thalidomide is also being used to treat diseases in which there is a massive growth of new blood vessels – diseases such as the cancers in which there is a solid tumour, diabetic retinopathy (a disease of the blood vessels of the retina caused by diabetes which leads to blindness) and macular degeneration. (The macula is that part of the retina that does not have blood vessels in front of it and which is used for fine-detail vision such as threading a needle.) Thalidomide is useful here, because it is a potent inhibitor of the growth of new blood vessels.

Living creatures are made from left-handed amino acids. Proteins are left-handed. Meat is made of proteins, and proteins are made of 20 or so amino acids. Glycine, one of these amino acids, is the only one that is symmetrical (like a ball), and so doesn't have a left-handed form and a right-handed form. The other 19 amino acids can exist in left-

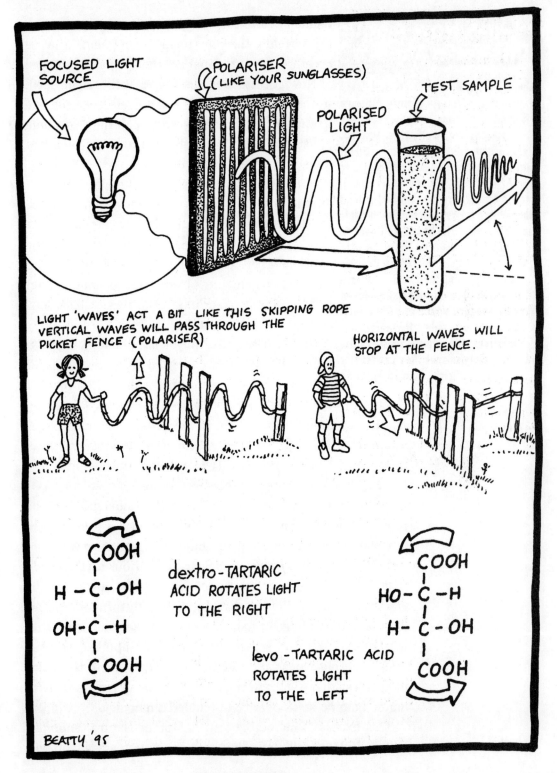

FOCUSED LIGHT SOURCE

POLARISER (LIKE YOUR SUNGLASSES)

POLARISED LIGHT

TEST SAMPLE

LIGHT 'WAVES' ACT A BIT LIKE THIS SKIPPING ROPE VERTICAL WAVES WILL PASS THROUGH THE PICKET FENCE (POLARISER)

HORIZONTAL WAVES WILL STOP AT THE FENCE.

dextro-TARTARIC ACID ROTATES LIGHT TO THE RIGHT

levo-TARTARIC ACID ROTATES LIGHT TO THE LEFT

BEATTY '95

And How About This for a Theory?

One scientist, Dr Dick Jeffries, a fossil expert at the Natural History Museum in London, reckons he knows why we humans are mostly right-handed. He reckons it's all due to an accident that happened about 500 million years ago. At that time, there was an animal called Cothurnocystis that had a head shaped like a boot and a primitive left ear in its anus. The animal fell over to one side, and it preferred to stay there. He reckons that it fell over on to its right side, and somehow as a result, we humans are right-handed some 500 million years later! But why, you might ask, didn't this affect our relatives, the chimpanzees and the apes so that they are mostly right-handed as well?

handed or right-handed forms in the laboratory – but virtually the only forms ever found in living creatures on our planet are the left-handed forms!

Now, it turns out that the universe, right down to the sub-atomic level, has a preference for one hand over the other – and according to some physicists, this asymmetry led to the proteins of life all being left-handed!

For a long time, physicists thought that sub-atomic reactions were completely symmetrical. Electrons have a spin, and physicists thought that in every sub-atomic reaction, you would have equal numbers of left-spinning or right-spinning electrons being involved. (Left-spinning and right-spinning electrons are also called left-handed and right-handed electrons.)

But in 1957, the physicist Chien-Shiung Wu found that electrons emitted in some forms of radioactive decay were practically all left-handed electrons, with hardly any right-handed electrons. Soon after, another asymmetry was found in the sub-atomic universe. A strange sub-atomic particle called the antineutrino was discovered. The antineutrino (which has no mass – matter – but which always travels at the speed of light) exists only as a right-handed entity. With our current understanding of physics towards the end of the 20th century, we know only that these asymmetries happen, but we do not know why.

We do know, however, that because of various asymmetries that exist in the four forces of nature, the attraction of electrons to the nucleus of the atom is slightly different for the left-handed and right-handed electrons. Somehow, this incredibly weak and subtle difference is amplified, so that once the electrons and atoms have combined to make molecules, there will always be a very small excess of left-handed amino acids over right-handed amino acids!

This excess is incredibly tiny. If Mother Nature randomly makes 200 000 million million and one amino acids, you will get 100 000 million million right-handed amino acids, and 100 000 million million *and one* left-handed amino acids. But back in the dim distant past, this tiny excess might have been large enough to create the world in which we live, where life as we know it can be made only from left-handed amino acids.

We still don't have all the answers yet, but perhaps we should leave the last word to the English metaphysical poet John Donne who in 1607 wrote in a letter to the Countess of Bedford: 'Reason is our Soul's left hand, Faith her right. By these we reach divinity.'

References

New Scientist, No. 1313, 8 July 1982, 'Lefthanders are born, not made' by Mary Gribbin, pp.100–101.

New Scientist, No. 1651, 11 February 1989, 'The left and right of brains at work' by Lesley Rogers, pp.56–59.

New Scientist, No. 1789, 5 October 1991, 'Copycat marmosets get a hand from their mums' by Clare Putnam, p.14.

Sydney Morning Herald, 9 December 1991, 'Line on life's left-handers' by Steve Connor, p.13.

New Scientist, No. 1916, 12 March 1994, 'Sudden death for left-handers' by William Bown, p.16.

Smithsonian, December 1994, 'Life for lefties: from annoying to downright risky' by Nancy Shute, pp.130–143.

Journal Watch, Vol. 13, No. 12, 'Thalidomide as therapy' by Al Komaroff, p.96.

Sydney Morning Herald, 3 July 1995, 'Twins mirror nature's winning hand' by Gavin Gilchrist, pp.1, 6.

HOLLOW ATOMS

SCIENCE HAS COME a long way in the last 2000 years, and now, in the late 1900s, not only can we make *hollow* atoms, we can even find uses for them – such as playing music! Thousands of years ago, some of the ancient Greeks began to think about atoms (which they described as small particles that could not be divided any further). The Greek philosophers Leucippus and Democritus thought that if you could know the movement of atoms through empty space, then you could understand all the events that happen in the real world. In fact, Democritus said, '...colour exists by convention, sweet by convention, bitter by convention, in reality nothing exists but atoms and the void'.

Atoms are the smallest building blocks of matter. In fact, the Greek word *atomos* means 'indivisible', or unable to be divided. There are over 90 different types of atoms in the world around us – starting off with the smallest and lightest atom of all, hydrogen, and working up through middle-weight atoms like iron, before reaching the heavy atoms like uranium and plutonium.

It was only in the early 20th century that we began to work out what an atom looks like. It looks a little bit like our own solar system.

At the centre of our solar system is the Sun. Similarly, right at the very centre of an atom is the nucleus. It's very small – about one-ten thousandth the diameter of the atom. Even though it's so small, the nucleus has most of the mass (the amount of matter) of the atom – about 99.98 per cent.

Electron

The electron was discovered by Joseph John Thompson in 1897. We've all seen the glow of a neon tube. Thompson made a primitive neon tube by filling a glass tube with a gas at a very low pressure, and then passing electricity through this gas. He saw a blue glow at one end of the tube. Through a series of experiments Thompson showed that this glow was caused by tiny negatively charged particles called electrons.

(Compared to our solar system, the Sun is very small, and it has most of the mass.)

In our solar system there are nine known planets orbiting around the central sun. In our model of the atom, there can be anywhere from one to over 100 electrons orbiting around the central nucleus. For example, uranium has 92 electrons.

Finally, our sun attracts the planets, thanks to its enormous gravity. In the atom, the nucleus is positively charged, and it attracts the negatively charged electrons.

But there are a few differences between the atom and our solar system – besides the obvious difference of size.

First of all, in our solar system, the planets have well-defined surfaces, and actually exist in just one part of their orbit at any given time. But in the atom, the electrons are not solid objects – they're sort of fuzzy, and don't actually exist in just one spot, but rather have a probability of being somewhere in their orbit.

Secondly, in our solar system there's only one planet per orbit. But in an atom, you can have more than one electron in each orbit. You can have two electrons in the innermost orbit, closest to the nucleus, and then eight electrons in the next orbit, and then eighteen, etc.

Thirdly, in our solar system, the orbits of the inner planets are fairly close to each other. As you move out, the orbits of the planets get further apart. But it's the other way around in the atom. The inner orbits are further apart, while the outer orbits can be quite close to each other.

Finally, our sun is made up of super-heated atoms of mostly hydrogen and helium, with much smaller amounts of lithium, boron, carbon and other elements. But the nucleus of the

Proton

Protons are very heavy (about 1826 times heavier than an electron) and have a positive charge.

In 1909, Hans Geiger and Ernest Marsden set up an experiment in which they fired alpha particles (the nuclei of the helium atom, the second lightest element) at a sheet of gold foil. Most of the alpha particles passed straight through the gold foil. But some of the alpha particles bounced straight back!

The New Zealand physicist Ernest Rutherford interpreted this experiment to mean that the structure of atoms (which was as yet unknown) was 'lumpy'. In 1911, he wrote: 'It was quite the most incredible event that has ever happened to me in my life. It was almost as incredible as if you fired a 15 inch shell at a piece of tissue paper, and it came back and hit you'.

The alpha particles were positively charged nuclei of helium atoms. Most of the time, they would pass through the gold foil, missing the positively charged nuclei of the gold atoms. But every now and then, one alpha particle would come very close to a gold nucleus, and would be reflected back.

What this meant was that the mass of each atom of gold was not spread evenly, but rather was concentrated in little lumps, some of which had the same sign of charge (positive) as alpha particles – and these positive paticles were protons.

Neutron

atom is made up of only two types of 'particles' – positively charged particles called protons, and neutral particles (with no charge at all) called neutrons. (There is one exception to this general rule – hydrogen, which has only a single proton in its nucleus.)

So to complete the picture, the solar-system model of the atom is that most of the mass is in the nucleus, which is made up of protons and neutrons. The heavier the atom, the more protons and neutrons it has. The nucleus has a positive charge because of the positively charged protons. Spinning around this central, very heavy, positively charged core are very light negatively charged particles called electrons. There are as many electrons as there are protons. These electrons move in a set of orbits.

In general, it's fairly easy to remove one or two electrons from the outer shells or orbits, but it gets harder as you try to remove more and more electrons (see box 'Pulling Off Electrons'). Usually, the electrons are removed only from the outer shells.

But recently, some physicists have done the surprising feat of 'making' atoms where the outer shells are full of electrons, but the inner shells close to the nucleus are empty – in other words, they have made a 'hollow atom'. (It's not really hollow, because the nucleus is still there, but the nucleus is very small, so they call it a hollow atom.)

So far, there are two main ways to 'make' a hollow atom. Either start with a naked nucleus and add electrons to it, or start with about 100 normal atoms and zap them with incredibly intense light.

The **first method** begins with a machine invented in 1987 by Mort Levine and Ross Marrs from the Lawrence Livermore National Laboratory in California. Their machine, the Electron Beam Ion Trap, could strip electrons from an atom. In 1994, they actually stripped a uranium atom of all of its 92 electrons, leaving behind its bare nucleus of 92 protons and over 140 neutrons.

This naked nucleus of positively charged uranium is

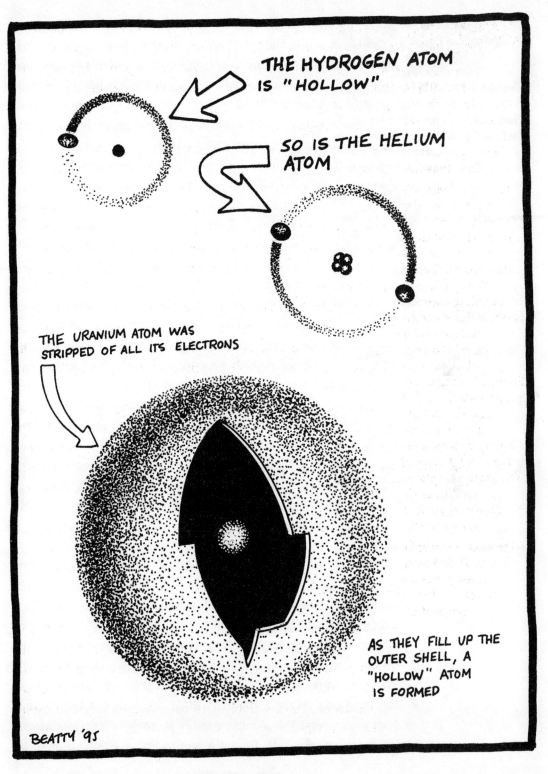

THE HYDROGEN ATOM
IS "HOLLOW"

SO IS THE HELIUM
ATOM

THE URANIUM ATOM WAS
STRIPPED OF ALL ITS ELECTRONS

AS THEY FILL UP THE
OUTER SHELL, A
"HOLLOW" ATOM
IS FORMED

BEATTY '95

Star Wars X-ray Laser

The Star Wars project (actually called the SDI, or Strategic Defence Initiative) was started by former US president Ronald Reagan. Its main aim was to stop enemy (read Soviet) missiles. One weapon was designed to stop these missiles while they were in space, with a laser beam. But it was not a beam of laser *light*, but of laser *X-rays!*

The X-ray laser, which was to be mounted on a satellite in orbit, was not your conventional laser. It was actually a nuclear weapon surrounded by rods of metal – like spikes sticking out from a central porcupine.

When the nuclear weapon was detonated, the expanding wave of radiation would vaporise the metal rods. But before they were completely destroyed, they would give off blasts of powerful, coherent, laser X-ray radiation, in the direction that the metal rod was aimed at. Of course, a squillionth of a second later, the expanding fire ball from the nuclear weapon would have destroyed the rods – but hopefully, the X-ray laser beams would, by then, be on their way to destroy enemy missiles.

incredibly thirsty for electrons. (In a normal atom, the number of negative electrons has to be the same as the number of positive protons, so that the normal atom is neutral, with no overall charge.) The first electrons that it captures fill the outer electron orbits because these are the first ones they 'see'. So you end up with an atom, in which most or all of the electrons are in the outer orbits. You now have your 'hollow atom'.

Of course, this hollow atom is very unstable and it immediately begins to revert to a normal atom. The negatively charged electrons head for the positively charged nucleus, but they can't get there in one hop. They have to skip down through each electron shell (the potential place where one or more electrons can live), and as they drop down, they give off energy. And this all takes time.

This hollow atom is desperately hungry for electrons. It will 'suck' any electrons that it comes across onto itself. Therefore, this hollow atom is loaded with potential energy. If it still has some unfilled inner shells when it happens to run into a target, it will give off this potential energy into the target (because the target is full of electrons).

Dieter Schneider and his fellow workers at Lawrence Livermore blasted hollow atoms of uranium and xenon onto surfaces of mica. (Mica is the name of a group of minerals that can be easily cleaved into flexible sheets, which can be used as insulators, or as windows in older domestic coke stoves.) The energy from the hollow atoms made tiny blisters just a few billionths of a metre across on the surface of the mica. These blisters are just like the little blisters or pits that you find in a CD. But they have one-ten thousandth of the area of the standard pits on the CD or CD-ROM that you can buy today. This means that you could possibly store 10 000 times more information on a CD, than we can today. A CD with such a high information storage density would be useful.

The **second method** of making hollow atoms uses very bright light, and this also could lead to a new useful tool for our society – an X-ray laser.

This work was done by Charles Rhodes and his colleagues from the University of Illinois in Chicago. They blasted a very intense light from an ultra-powerful ultraviolet laser at a group of about 100 atoms. This spot of light was incredibly small (only three-millionths of a metre across) and incredibly bright. In fact, the pulses from the ultraviolet laser had a peak power of some 800 gigawatts – that's over 10 times the total electrical generating capacity of the United Kingdom! Of course, this power was delivered for a very short time – only 0.3 million-millionths of a second.

We still don't know exactly what happens. Normally, in a group of atoms, the electrons belonging to each atom are quite tightly bound to that atom. But Charles Rhodes thinks that all the outer electrons of the 100 or so atoms seem to move as a group, and all swing from one side of the group of atoms to the other side, and back again. Somehow, and the physicists still don't know how, the combined energy of these oscillating electrons 'focuses' on to the inner electrons, which are normally very tightly bound to the nucleus. This energy boots these electrons into higher orbits within the atom, to create our hollow atoms.

Of course, these hollow atoms are unstable and try to turn back into normal atoms by giving off radiation. (The only way that an electron can drop from one orbit to the next is by giving off energy. This energy can take the form of one of the many types of radiation – visible light, infra-red radiation, X-rays, or even gamma rays.) In this case, they give off X-rays. This is very exciting, because it could lead to the first

Pulling Off Electrons is Hard to Do

An atom is electrically neutral. It may have (say) 92 positively charged protons, but it also has 92 negatively charged electrons. When we remove one electron from an outer shell or orbit, there are still 92 protons, but only 91 electrons. So now the atom has a net positive charge of 1.

When we try to remove the second electron, we find that the job is harder. We are trying to pull a negatively charged electron away from an atom which has a net positive charge. Opposite charges (positive and negative) attract, and the positively charged atom has a strong pull on the negatively charged electron.

As we remove more and more electrons, we increase the amount of positive charge on the atom. This increasing positive charge makes it harder to pull away additional electrons. This has always been the major stumbling block in removing all or most of the electrons from an atom.

Atoms

Atoms are very small. A typical atom is about one-tenth of a billionth of a metre across. A hydrogen atom weighs about one million million million millionth of a gram. A single drop of water is made up of over a thousand million million million atoms.

practical X-ray laser.

Now the Star Wars project did produce an X-ray laser, but because the basis for its operation was a nuclear weapon, it wasn't very practical (or desirable)! But Charles Rhodes' sensible X-ray laser could be very useful.

Firstly, because X-rays have a shorter wavelength than light, they could actually look at things too small to be seen by light microscopes, such as proteins and viruses. They could show us much more detail than we have ever seen so far.

Secondly, because X-rays can penetrate living tissue, they could actually examine tissue while it is still alive. At the moment, practically all our work with microscopes is done on dead tissue.

And finally, these yet-to-be-invented X-ray lasers, with their very short wavelengths, could read the very densely packed information in the yet-to-be-invented CDs that would also be made using hollow atoms.

At the moment, these hollow atoms are just a laboratory tool. But soon we could be finding them, and their spin-offs, in our cars and our shops, and even in our toys.

References

The Academic American Encyclopedia (electronic version), Grolier, Inc., Danbury, CT.

Microsoft ® Encarta. Copyright © 1994 Microsoft Corporation. Copyright © 1994 Funk & Wagnall's Corporation. (CD-ROM).

New Scientist, No. 1980, 3 June 1995, 'Rebuilding the atom' by Ian Hughes and Ian Williams, pp.31–33.

New Scientist, No. 1985, 'Heart of the atom' by Christine Sutton, Inside Science, pp.1–4.

GROWING SKIN AND ORGANS

ABOUT 10 YEARS AGO, scientists began to grow sheets of skin in their laboratories. Today, in the mid-1990s, a Californian biotechnology company called Advanced Tissue Sciences is gearing up for full production of sheets of artificial human skin, roughly the size of the palm of your hand. They call this technology 'Tissue Engineering', and in 10 years time, these and other biotechnologists want to start selling entire organs or tissues, like female breasts, or knee or hip joints.

There are two stages in trying to 'grow' body parts – first, to grow skin (which is easier), and second, to grow a complete organ (which is harder).

There are a few reasons why the biotechnologists started off with trying to grow skin. First, pathologists have long experience in growing flat sheets of cells. Second, skin is probably the easiest tissue to grow from the body. Third, there is a huge market for skin. In America alone, about 2 million people get burnt each year, of whom over 100 000 end up in hospital, and over 10 000 die. The worldwide annual market for artificial skin is around $300 million for burns victims, and $4000 million for patients with diabetic ulcers and intractable bedsores!

The skin, your largest single organ, weighs about 16 per cent of your body weight, and it is pretty complicated. It has special receptors for pressure, vibration, light touch and temperature, as well as naked nerve endings that respond to heat and pain. Skin is a chemical and mechanical barrier

A Brief History of the Epidermis

There are several layers of cells in the epidermis. As they migrate upward, they change shape from columns, to spheres, to flat cobblestones. It takes about 27 days for the full cycle to occur.

The columnar cells in the basal (or bottom) layer divide into new cells every day, to replace the cells lost from the surface. This takes a lot of energy, so they usually do this when the other demands upon the body's energy supplies are low – about 4 am!

On average, during your entire lifetime, you will shed a total of about 18 kilograms of skin.

Early Transplants

A 15th century oil painting by Alonso de Sedano shows a leg transplant operation. Saints Cosmos and Damian, who are the patron saints of physicians and surgeons, have removed a cancerous leg from a white man and replaced it with a black leg from a Moor.

In 1665, Samuel Pepy's diary talks about an experiment at Gresham College. Blood was transfused from one dog to another. The second dog survived quite well. Nowadays, when you go to the vet to get a blood transfusion for your anaemic cat, they don't take a little bit of blood out of a whole lot of cats. Instead, the blood for your cat is drained entirely out of a single cat (one whose life is obviously considered expendable).

In June 1902, a Dr Emerich Ullmann of Vienna transplanted the kidney of a dog into the neck of a goat – and the kidney and the goat survived long enough to be presented at a meeting of the local surgical society. It was the first report of a kidney transplant, and the last report by the good Dr Ullmann, whose name never appears in the medical literature again.

which stops you from leaking out all over the place, and which stops outside poisons from getting inside your body and contaminating your precious bodily fluids.

Each square centimetre of your skin has about 3 million cells, 100 sweat glands, 4 metres of nerves, 3000 sensory cells, 150 nerve endings, 10 hairs and hair follicles, 12 sebaceous glands and about 1 metre of blood vessels.

Your skin is made from two quite separate layers. The outer layer (the epidermis) is quite thin. It ranges from a tenth of a millimetre over the eyelids, to up to 5 millimetres thick on the soles of your feet. The cells in your epidermis are supported and nourished by the next layer underneath, the dermis. The inner layer (the dermis) is a thick, viscous gel of collagen with various useful things standing around in the gel – blood vessels, lymphatic vessels, nerves, sweat glands, oil glands and hair follicles.

Trying to 'make' something as complicated as skin is about as difficult as trying to make a corn plant from scratch. But luckily, farmers don't have to 'make' a corn plant – they just put corn seeds in the right kind of soil. And likewise the biotechnologists from Advanced Tissue Sciences found that they could plant young skin cells as seeds. Instead of soil, they used an inert scaffold of polylactic acid and polyglycolic acid – the same biodegradable chemical polymers that are used in those surgical stitches that just dissolve away.

In 1992, Joseph Vacanti, a transplant surgeon working at the Harvard Medical School, and Robert Langer, a chemical engineer from the Massachusetts Institute of Technology, began seeding their inert polymer scaffolds with healthy young dermis cells. They got their dermis cells (their seed cells) from the surgically removed foreskins of circumcised baby boys. In their factory, they can culture from one single foreskin enough dermis (the inner skin layer) to cover six football fields!

One of their clinical trials involved deep foot ulcers in diabetic patients. After they had laid their factory-grown sheets of dermis over the ulcers, blood vessels from deeper layers began to grow into the new dermis. Over the period of a year, the factory-grown sheets of dermis merged into the patients' own dermis. Even better, the patients' own epidermis began to migrate on top of the factory-grown dermis. In a trial involving 400 patients, they achieved a 50 per cent cure rate – much better than the 8 per cent cure rate achieved when using traditional methods.

Another company, Genzyme Tissue Repair, also in Cambridge, Massachusetts, has taken a different approach. They grow sheets of the outer layer, the epidermis. If you give them a month's time, and some healthy skin cells from your patient, they'll grow a sheet of fully compatible skin, one quarter the size of the palm of your hand, for $350. The average burns victim will use up over $60 000 worth of artificial skin.

It seems that we are getting near to being able to grow skin as it's needed. But now, in the mid-1990s, we are also getting closer to the second stage of 'growing' body parts – i.e., growing a complete organ. Believe it or not, various biotechnologists have already been trying to 'grow' a three-dimensional organ (such as the female breast) in the laboratory!

There is a need for a female breast that could be 'grown' as required. Each year, worldwide, there are some quarter of a million mastectomies, where a breast, usually attacked by cancer, is surgically removed. Many women think about whether they want to replace the removed breast. There have been various silicon implants and salt water implants tried –

Transplants and Rejection

Many skin grafts were tried on burnt airmen during World War II. However, most of them failed, because the new skin (from a volunteer) was rejected by the immune system of the burnt airman. However, if the airman was lucky enough to have an identical twin, the skin graft was much more likely to survive.

Your immune system has the job of recognising and repelling invaders. There are various chemical 'markers' on the outside of our cells, and our immune system usually recognises these as acceptable, and most other markers as foreign. So if you culture your own cells to make a transplant organ, the organ will be accepted by your immune system.

One advantage of using the dermis in skin transplants is that some of the cells in the dermis, the fibroblasts, tend to have fewer skin markers (in fact, they have no surface markers in the class called 'MHC Class 2 Antigens'). This means they are less likely to be rejected.

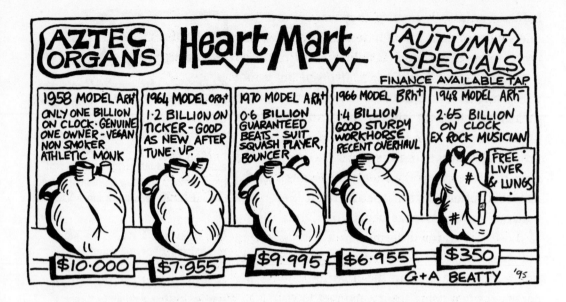

but they have met with varying degrees of success. Well, women may soon have a better option than conventional surgery.

Reconstructive surgery has an ancient and honourable tradition. In fact, it's one of the older forms of surgery. Four thousand years ago, in India, law-breakers were punished by having their noses chopped off. And so, 2000 years before the birth of Christ, the Caste of Potters invented a nose-reconstruction technique, which is still used today. They rebuilt the nose by using part of the forehead!

But nowadays, all too often the surgeons have to use artificial materials – like wood for legs, silicon implants for breasts, or Teflon and titanium for hip joints. If only the surgeons could get hold of enough natural tissues or organs to use in reconstructions, their job would be a lot easier. Two-dimensional sheets of skin are relatively straightforward to grow, but when you try to 'grow' a three-dimensional organ such as the female breast, you run into a whole different bunch of problems.

The female breast is quite a complicated organ. It's basically a modified sweat gland that can turn blood into milk. Even though a single breast weighs about 170 grams, it can deliver up to $1\frac{1}{4}$ litres of milk per day.

A breast is made up of about 20 smaller mammary glands, each of which can make milk. Each of these small mammary glands has a separate duct or pipe (about 3 centimetres long) which opens out onto the nipple. Each of the 20 mammary glands, with its duct, looks a little like a tiny tomato bush with thousands of tiny tomatoes on it. The milk is produced in the tiny bubbles which look like tomatoes, and travels along the stem until it gets to the nipple. At the nipple, the opening of each duct is almost too small to be seen, about 0.5 millimetres in diameter.

These 20 mammary ducts are not just straight pipes, however. Just before each mammary duct gets to the nipple, it has a bulge or swelling, where milk collects. So as soon as babies start sucking on the breast, they get an instant feed from the milk stored in the swelling. But the baby doesn't have to do all the work. There are muscle fibres arranged in a circular pattern, just underneath the skin of the nipple. When these muscles contract, they squeeze the milk out into the baby's mouth. The milk can squirt distances of up to one metre!

Trying to make something as sophisticated as a breast is virtually impossible with science and technology as it stands today. For one thing, there's lots of fat cells in the normal female breast, and in the mid-1990s, we simply don't know much about growing fat cells in the laboratory.

Now David Mooney, a chemical engineer from the University of Michigan, has been working with researchers from the Carolina Medical Center in Charlotte for the last seven years. Together, they've been trying to develop a *laboratory*-grown female breast. Today, it's much too hard to make something that does everything that a real female breast can do. So the team has chosen an easier goal – they've been trying to make something that *looks* very much like a female breast.

They start off by using a fine-meshed synthetic honeycomb scaffold. They want to seed this mesh in different areas, with

Jasmine Flowers Stop Breast Milk

The average Australian baby is breast-fed for around four months. In that time, the baby drinks about 100 kilograms of milk, which is more than the weight of the mother, and doubles his or her average birth weight of 3.5 kilograms.

But there are certain occasions after a baby has been born when you don't want to start breast-feeding. For example, the new-born baby might be given up for adoption, or there could be a stillbirth (when the baby is born dead), or the baby could die within 24 hours of birth.

When this happens, the breasts are primed to deliver milk, but there is no baby to take the milk. Normally, it takes about three days after the baby is born for the milk to come in. Then there's a period of about four days when the breasts get very full and uncomfortable, until the supply of milk comes into balance with the amount that the baby is drinking. But if baby is not there, not only does the mother have to cope with the baby's absence, she has to deal with painful, engorged breasts.

One treatment to stop milk production is a drug called bromocriptine, which should be started as soon as possible after the baby is born. Unfortunately, bromocriptine has side effects such as nausea, dizziness and an increase in the production of milk after the drug is stopped. The dizziness is usually managed by the woman taking the bromocriptine first thing in the morning before a long sit-down breakfast, or last thing at night, just before going to bed. There has been no real cure for the unwanted increase in milk production after taking the bromocriptine, except to ride it out, and put up with the discomfort.

But now there's a brand new herbal treatment, one that's been used for thousands of years in the East! In India, the fresh flowers of the Jasmine vine are often strung onto a fine cotton thread, and used to beautify the hair. But in southern India, they also have a medical use — they're used to stop the unwanted production of breast milk. The traditional way is to buy two strings of jasmine flowers 50 centimetres long. A string of flowers is simply laid over a breast, and stuck down with sticky tape. The women wear loose blouses over the flowers, without a bra, and replace the flowers every 24 hours for five days.

This ancient herbal remedy was put to the test at the Department of Obstetrics at the Christian Medical College Hospital, Vellore, in India. The study was carried out with 60 women who needed to have their breast milk stopped — either because of a stillbirth, or the death of a baby within 24 hours of birth. They were divided into two groups, one to receive the expensive bromocriptine treatment, the other to receive the traditional jasmine flowers.

They were followed up for two weeks. The treatments were ➤

▷ compared by two yardsticks. The first yardstick was milk production, which was rated from no milk, to a few drops of milk with breast pressure, to a moderate amount of milk with pressure, to lots of milk with no pressure. The other yardstick was breast engorgement, which was rated from no engorgement, to slight engorgement, to moderate engorgement with no painkillers needed, to extreme engorgement accompanied by severe pain that required painkillers.

It was a narrow win for the ancient herbal medicine. Jasmine flowers gave the same results as bromocriptine, but without the expense or the side effects, and with the advantage of a nice smell.

At the moment, they don't really know exactly why jasmine flowers work in stopping milk production. They do know that in breast-feeding rats, the smell of jasmine flowers works fairly well, but that the effect is much stronger if the jasmine flowers are applied directly to the rats' breasts/mammary glands. So there's probably some unknown hormonal effect set off by the lovely odour of the flowers going up through the nose and setting off circuits in the brain, and there's probably also an unknown local effect on the breasts. On the other hand, they use cold cabbage leaves in the post-natal ward of a Melbourne hospital, and they also claim good results, even though they haven't been tested in a formal experiment.

The jasmine flowers will probably help the discomfort if you want to stop breast-feeding after a few months. There's only one problem – in Australia, jasmine flowers are available only in spring, and nobody's done the research to see if jasmine tea has the same effect.

different cells. These cells would be stimulated into growth by chemicals called 'growth factors'. Instead of using fat cells (which they don't understand), they've been working with cells that they do understand – smooth muscle cells, which are usually found lining the gut. En masse, smooth muscle cells feel resilient and elastic, so they'd be quite acceptable in a replacement breast. Now there is the problem that smooth muscle cells are continually moving – they're always contracting and relaxing. But Mooney reckons that his team can solve this.

There's another big problem, however. So far, the only way they can 'grow' cells is as a thin surface layer. They can't yet grow cells deep *inside* a tissue, away from the blood vessels that deliver oxygen and food and remove waste products. Another problem is that the blood vessels that they can 'grow' look good under a microscope, but when they pass blood through them, they simply burst at the seams!

Even so, Mooney reckons that by the year 2000, they'll have laboratory-grown female breasts available for the surgeons to use in human trials.

With a little more work, the boffins should be able to flesh out their ideas, and come up with something a little more substantial.

References

Australian and New Zealand Journal of Obstetrics and Gynaecology, Vol. 28, 1988, 'Suppression of puerperal lactation using jasmine flowers (Jasminum Sambac)' by Pankaj Shrivastav, Korula George, N. Balasubramaniam, N. Padmini Jasper, Molly Thomas and A.S. Kanagasabhapathy, pp.68–71.

New Scientist, No. 1968, 11 March 1995, 'Life in the tissue factory' by Steven Dickman and Gabrielle Strobel, pp.32–37.

Sydney Morning Herald, 30 March 1995, 'Foreskins: an expanding industry' by Jill Margo, p.11.

FINGERPRINTS

SHERLOCK HOLMES WAS able to finger crooks by examining their footprints, though nowadays, the police use *fingerprints*, not *footprints*, to track down criminals. But it seems fingerprints might also give the medical profession some vital clues – about your future diseases!

The famous American gangster John Dillinger burnt his fingerprints off with acid – but he could have saved himself the intense pain, because they grew back. John Phillips, another gangster, had new fingerprints grafted onto his finger pads – but the police convicted him because they got a match with the prints further down his fingers.

The pads on your fingers are not perfectly smooth, but have tiny ridges and valleys on them – to increase the friction, and give you a better grip. These ridges and valleys happen because of the way that the outer layer of your skin (the epidermis) and the inner layer (the dermis) lock into each other – like a zipper. Ridges on the top of the dermis lock into valleys on the bottom of the epidermis, and carry through as a ridge on the top of the epidermis. When you wrap your hand around a glass, and then remove it, you can see the impression of those ridges on the glass – your fingerprints.

Fingerprints have been used for identification for thousands of years. The ancient Chinese and Assyrians used fingerprints on legal documents. The ancient Kings of Babylon would press their entire right hand into a slab of clay carrying a decree, before it was fired. Even back then, it was a common belief that no two people shared the same fingerprint.

The Hardest Fingerprint to Find

According to *Biometric Technology Today*, a young female South-East Asian bricklayer would be *'the fingerprint specialists' nightmare'* !

Young people have softer skin, so their prints are worn off faster. *Females* tend to have finer fingerprints than males. *South-East Asians* also tend to have the finest fingerprints of any people on our planet. And *bricklayer*? Bricks are very rough, and they tend to rub the print off the finger, or at least, wear the ridges fairly flat.

Thomas Bewick, who lived from 1753 to 1828, was a master engraver of wood blocks. He knew that fingerprints were unique, and he used his thumbprint to identify his work.

In 1823, the Czech physiologist Johannes Evangelista Purkinje was studying sweat glands in fingers. These sweat glands discharge their sweat into the grooves between the ridges of your fingerprints. He soon noticed that every fingerprint that he examined was different and suggested that fingerprints could therefore be used for identification. (Purkinje was a very lateral scientist – he did ground-breaking research into the organs of the brain, eye and heart, as well as general research in the fields of pharmacology and embryology. He has named after him the Purkinje cells of the cerebellum, the Purkinje fibres of the heart, and the Purkinje germinal vesicle of the human egg.)

In 1858, in India, the British magistrate William Herschel would get Indians to 'sign' their contracts with the print of their entire hand, to make them more aware of the significance of the document. He soon realised that the fingerprint was also an effective way to identify someone, and began to concentrate on fingerprints, rather than handprints. He noticed two things. First, every fingerprint he found was different, and second, fingerprints did not change as a person grew older.

But it was a Scottish doctor working in Japan, Henry Faulds, who became the first person to solve a crime with fingerprints. He matched the fingerprints found on a cup at a robbery in Tokyo with those of a servant, and described his feat in a letter to the prestigious science journal *Nature* in 1880. When William Herschel returned to the UK from India, he read the report by Henry Faulds, and he also wrote to *Nature*, to describe his own work in this field. These two letters were largely ignored in Europe for several years.

Quite independently, however, Juan Vucetich, an Argentinian police officer in Buenos Aires, became interested in fingerprints. In fact, he was the first person to regularly take fingerprints in ink.

Vucetich soon realised that the big problem was not how to take fingerprints, but how to classify them so that other police officers could match a fingerprint found at the scene of a crime with a previously stored fingerprint. In 1888, he published a paper on how to analyse and classify fingerprints so that they could be easily filed and later retrieved for matching.

Four years later, in July 1892, this information was used to solve a crime. An Argentinian police officer, Inspector Alvarez, was investigating a crime. In the nearby coastal town of Necochea, the two children of Francisca Rojas had been beaten to death. Inspector Alvarez found a bloody thumb print on the door, so he sawed out that section of the door and took it along to the police station. He then had the children's mother fingerprinted, and when she learnt that her own thumb print matched the thumb print on the door, she confessed to the crime. Shortly after, Argentina became the first country in the world to use fingerprints as the chief method of identifying offenders.

The British Empire was a little slower to adopt fingerprints in the pursuit of crime.

In 1888 (the same year that Vucetich published his paper), Sir Francis Galton, who was a cousin of Charles Darwin, remembered the two letters published in *Nature* some years before. He soon realised that fingerprints were a good way to identify criminals uniquely.

In 1896, Edward Richard Henry, the inspector general of the Bengali Police Force in India, invented another way of classifying prints – a method that was quite different from the one that Vucetich proposed. By 1901, Galton and Henry had teamed up, and came up with the Galton-Henry system of fingerprint classification. Scotland Yard put Galton in charge of the Criminal Investigation Department, and in the first six months alone, his Fingerprints Branch successfully identified over 100 criminals. Even though Henry later became Sir Edward Richard Henry, he was always called 'Mr Fingertips'.

The Galton-Henry fingerprint classification method is used

Genetics and Fingerprints

There are a few genetic differences in people's fingerprints. Pygmies, the Bushmen of Africa, and some central Europeans tend to have the most arches. The Ainu people in Japan, Europeans and black Africans tend to have more loops than other people, while Mongolians and Australian Aborigines tend to have more whorls than anybody else. This does not mean that Mongolians and Australian Aborigines are more likely to have high blood pressure – it just means that they are more likely to have whorls. Nobody has yet done the research to see if Australian Aborigines with more whorls have higher blood pressure than Australian Aborigines with fewer whorls.

In fact, there is also a weak relationship between the different patterns on your fingerprints and your blood group (which depends on your 'racial' group).

throughout the entire world, apart from South America and China, where the Argentinian classification system is still used.

There are four basic patterns used to classify fingerprints – the **arch**, the **loop**, **whorl**, and the **composite** (which is a combination of the arch, loop and whorl). Some of your finger pads have a point called a 'triradius', where three sets of ridges meet. If you have an arch on one finger, that means that there is no triradius. A loop is where you have one triradius, while a whorl is where you have two triradiuses.

Another way to picture them is to imagine that you're looking at a map with contour lines to show the hills and valleys – like a military map or a bushwalker's map. An 'arch' is a gentle rise, a 'loop' is a ridge, while a 'whorl' is a solitary hill or peak.

The fingerprints that you leave behind on all the surfaces that you touch are actually patterns of the incredibly tiny amounts of moisture from the sweat glands on your fingertips. Surfaces that are non-absorbent (like glossy paint or glass) leave a good fingerprint, while absorbent surfaces like fabrics tend to leave lousy fingerprints. In fact, fingerprints are usually invisible unless you've rolled your fingers in some blood or ink or paint. So the police will usually start off by dusting the surface with a very fine powder, such as aluminium, which sticks to the moisture.

There are now more sophisticated techniques to get fingerprints, though. Your sweat contains many chemicals, such as vitamins, amino acids and fats. A few of these chemicals will fluoresce (like your white shirt under a black UV light). So forensic scientists have used a 5 watt laser to make fingerprints fluoresce. With this technique, they can get two-week-old fingerprints from paper which has been baked and then soaked, and even from cloth and human skin. The forensic scientists have also noticed that with this laser-fluorescence technique, that fingerprints that have been there for a long time are more orange, while younger fingerprints are more green. This gives the possibility of working out just

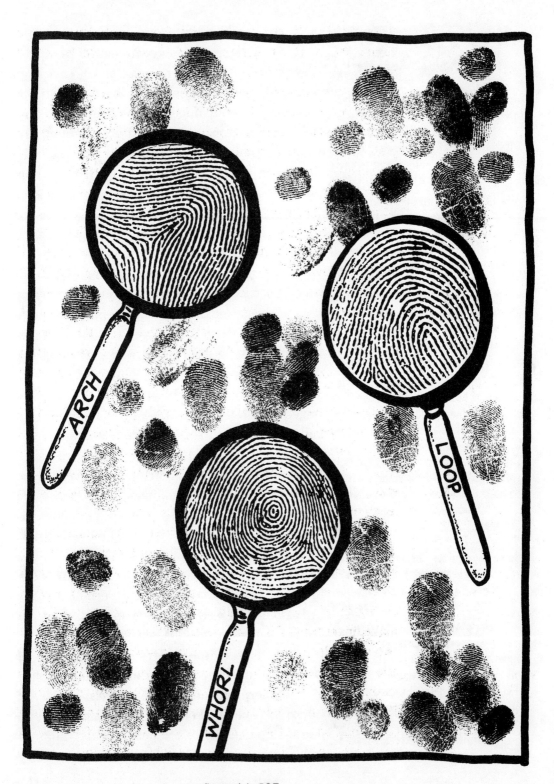

Fingerprints and the FBI

In the early part of the 20th century, in the USA, there were many different bodies collecting and using fingerprints. These organisations did not talk with each other, and so there was much inefficiency.

In 1924, a central organisation was set up to co-ordinate the use of fingerprints – the Identification Division of the Federal Bureau of Investigation. It began with 810 000 sets of fingerprints from the International Association of Police Chiefs and the Federal Penitentiary at Fort Leavenworth in Kansas. These fingerprints were from people who had been arrested.

In 1933, the Civil Identification Section was set up, to collect fingerprints from people who had not been arrested. As a result, the number of fingerprints held by the FBI increased enormously. There was another enormous increase during World War II, when most US government employees and workers in war plants had to give their fingerprints to the FBI. By the mid-1990s, the FBI have about 200 million sets of fingerprints (29 million of them from criminal suspects) on file – with about 60 000 new sets coming in every day!

One major problem is how to scan this enormous data base for the fingerprints that match the ones found at the scene of a crime. In 1995, the FBI awarded 18-month, multi-million dollar contracts to three companies, to come up with an automated fingerprint identification system. After the contracts have run their ➤

when the fingerprint was laid down.

David Harper and his co-workers of the Police Forensic Science Laboratory in the UK have even found a bacterium which loves to eat some of the chemicals in your fingertip sweat. They set up a deliberate search to find such a bacterium, and they screened over 600 different bacteria. The bacterium they finally settled upon, *Acinetobacter calciacaticus*, was one that they had swabbed from the forehead of a casual visitor to their laboratory! They now keep in their laboratory freeze-dried cultures of these bacteria, which will grow only where your fingertip sweat was. The bacteria will grow and join up the tiny dots of sweat, making the fingerprint much more visible. With these and other techniques, it has become possible even to get fingerprints off the inside of gloves.

To match one fingerprint to another fingerprint, the fingerprint officer has to find a certain number of recognisable features that are the same in each fingerprint – these are called similarities. In most countries, there have to be more than eight such similarities between the fingerprints before they can be counted as admissible evidence – but in some countries, there have to be 17 similarities.

But not only can the patterns of your fingerprints help in tracking down criminals, they can also have a medical significance.

Back in the early 1980s, a team lead by Dr C.M. Habibullah from the Gastroenterology Unit of Osmania University in Hyderabad in India looked at the fingerprints of some 150 adult males, 90 of whom had duodenal ulcers. They found that there was a slight tendency for people with whorls to have more duodenal ulcers. Another study looking at American Japanese in Hawaii showed that if men had more whorls on their fingers, they were more likely to have a heart attack.

It was a British study, though, which found that whorled fingerprints were associated with high blood pressure in adults. This study was done by Professor David Barker (Head of the Medical Research Council's Environmental Epidemiology Unit at Southampton General Hospital), Dr Keith Godfrey (a Medical Research Council training fellow), J. Pearce (a research assistant), J. Cloke (a senior fingerprint expert) and C. Osmond (a statistician). The team was looking to see if there was any link between the growth of the foetus in the mother, the blood pressure of the baby once it had grown into an adult, and the adult's fingerprints.

They were lucky to find that the Sharoe Green Hospital in Preston had kept unusually detailed and complete (for that time) birth records on all babies born from 1934 to 1943. The records included data such as the baby's weight, length, and head circumference, as well as the weight of the placenta, and even the mother's last menstrual period. In that period, some 1298 babies were born, and the team was able to track down some 450 of them.

They noticed that if babies were born very *thin*, they were more likely to have *whorls* on their fingertips, and to have *high blood pressure* when they became adults. (The average systolic blood pressure was 136 mm Hg if the subjects had no whorls on any fingers at all, compared to 144 mm Hg if they had a whorl on one or more fingers.)

The team is not exactly sure what's going on, but it is well known that fingerprints are laid down between the 13th and the 19th week of gestation

▷ 18 month period, a winner will be chosen. The winner will be expected to have a scanning-matching system running by 1998.

Sweaty Palms and Fingerprints

Czech physiologist Johannes Evangelista Purkinje realised that fingerprints were visible because of the sweat coming out of the tiny sweat gland ducts that open into the valleys between the ridges on your finger pads. But why have sweat there at all? Surely sweat would just make your grip more slippery?

Not so, according to David Robertshaw, a physiologist at the Cornell College of Veterinary Medicine. Dogs sweat from their footpads, so he did an experiment where he trotted a few dogs up an inclined treadmill, and measured how much they sweated, and how much friction grip they had. He and his colleagues found that the sweat actually increased the friction, and reduced the slipping. When he gave the dogs a drug to stop them sweating (atropine sulphate), the dogs began to slip more as they lost their grip. The sweat also made the skin on the footpads more sensitive, and more tough.

Fingerprints and Sex

A Japanese team of scientists can tell your sex from your fingerprints, even if you have long departed the scene!

The scientists are Tshizu and Yuji Yamamoto from the Department of Forensic Medicine at the University of Okayama. They look carefully in the actual fingerprints, and they find tiny skin cells left behind. Some of these skin cells will have your DNA inside. They duplicate the DNA so that they have lots of DNA to work with.

They then look for X and Y chromosomes. If they find an X and a Y, they know that a male left the fingerprint, but if they find two X chromosomes, they can blame a female for the fingerprint. Japanese police are interested in using this technique to help them in their investigations.

in the womb. According to the mathematicians, if the foetus has flattened finger pads, it's more likely to have the simpler arch pattern, or the slightly more complicated loop pattern. But on the other hand, if the foetus had swollen finger pads, it's more likely to have the more complex whorl pattern of ridges.

But why would a baby who was going to be born *thin*, have *fat* finger pads? One possible theory to explain this is that a foetus which does not have enough blood flow to feed its whole body might switch most of the available blood to feed its brain, and cut down the flow to the rest of its body. By feeding more blood to the brain, it will also accidentally feed more blood to the arms, and the fingers – which will appear as increased swelling in the finger pads. (The blood supply to the arms comes off the arteries which supply blood to the brain.) By cutting down the flow of blood to the rest of the body, it will starve the rest of the body of blood, and so the baby will be born with a thin body.

They're really not sure why there is a link between whorls and high blood pressure – but they do know the link is there.

Now, it's still early days in this research, so there's no point in getting a magnifying glass and looking at your fingertips and getting paranoid. You should be proud of your fingerprints – they're your very own personal ID, which you always carry with you. Various organisations try to collect fingerprints, but it's the FBI that has the biggest collection. They have over 200 million sets of fingerprints – but then, so does every house with small children!

References

New Scientist, No. 1312, 1 July 1982, 'Lasers help the mounties get their man', p.26.

New Scientist, No. 1354, 21 April 1983, 'Ulcer diagnosis at your fingertips', p.148.

New Scientist, No. 1563, 4 June 1987, 'Bacteria fool the light-fingered thief' by Ian Mason, p.40.

CORRUGATED ROADS

AUSTRALIA HAS ABOUT 900 000 kilometres of roads. Only about 50 per cent of these roads is paved. This puts us ahead of Turkey and Iceland (which have about 15 per cent of their roads paved), slightly behind Finland, New Zealand, Spain and the USA (which have about 55 per cent of their roads paved), and well behind Austria, Denmark, Italy and the UK (which claim to have 100 per cent of their roads paved). But it doesn't matter if your roads are paved or unpaved, they can all get corrugations – and the reason why was worked out way back in the early 1960s.

Now the oldest known road in the world is the so-called Sweet Track near Glastonbury, in the south-west of England. It's a wooden walkway that was built some 6000 years ago. It was discovered by a labourer, Raymond Sweet, while he was digging peat in a peat bog. He found a plank of a hard wood and brought it to the attention of archaeologists. They soon found a network of ancient wooden roads. They reckon that in this area, late in the Stone Age, people had lived in various villages that were separated by a reed-filled slushy swamp. The villagers had laid down timbers and planks across the swamp to make reliable all-weather roads, to connect their villages. One track was 1.8 kilometres long – not bad for 6000 years ago!

This same 6000-year-old technique of timbers-floating-on-swamp was used to build parts of the Alaska Highway. In June 1942, the Americans were panicking. It was less than seven months since Pearl Harbor, and the Japanese had already invaded American territory – three of the Aleutian Islands that zipper from the lower left-hand corner of Alaska out into the Northern Pacific Ocean. So the Americans urgently needed a road into Alaska, to protect it. This incredible road, that runs nearly 2500 kilometres across some five mountain ranges and over 120 rivers, was built in just 258 days by 16 000 people.

The road-builders by-passed the swamp whenever they could. And when they couldn't, they would 'corduroy' the road across the swamp – they put down many layers of pine trees, so the road just 'floated' across the swamp.

Two thousand years ago, the Romans knew that good roads were essential to help them govern their vast empire. By the time their empire collapsed, in the 4th century AD, they had built some 85 000 kilometres of roads. (For comparison, Norway today has about 85 000 kilometres of roads, Japan has about 1 100 000 kilometres of roads, while the USA has about 6 250 000 kilometres of roads.)

Roman road-building technology was not equalled for 1500 years. In the 19th century, Sir Robert Peel, the British prime minister, took six days to travel by coach from London to Rome – exactly the same time as the same journey had taken 17 centuries earlier, when the Roman Empire was at its peak! Their roads, and the services on them, were so reliable that civilians like St Paul could write a whole series of letters to his fellow Christians, secure in the knowledge that the letters would arrive.

The Roman roads lasted because of good design and construction. They were raised in the middle to let water run off to each side, and they had superb drainage on each side to carry the water away. Even today, the main road between Rome and Rimini uses a Roman tunnel built in 77 AD, more than 19 centuries ago.

The next great road builders were the Incas of South America, in the 1400s. Their two main roads were a coast road some 4000 kilometres long (the distance between Perth and Sydney), and another road in the Andean highlands some 5200 kilometres long. The highland road was the longest trunk road anywhere in the world until late in the 1800s. This road has tunnels that go straight through rock, steps that are cut into the sides of mountains, and suspension bridges flying over enormous chasms.

Altogether, the Incas built some 20 000 kilometres of roads. Like the Roman roads, these roads were used for communications, goods, and moving armies. At its peak, the Inca empire governed some 12 million people, who lived along most of the west coast of South America, and (to the east) in most of Bolivia and Argentina. The Incas did not use the wheel on their roads – everything was carried by llamas, or by humans. By having trained runners stationed every 3 kilometres along the highways, they could transport fish some 400 kilometres from the coast to the inland capital in just two days!

But today we use the wheel, and it turns out that there are three main ways in which the wheels of vehicles damage roads.

Firstly, the tyres gradually rub off the top layer of the road, making it more smooth. This leads to skidding, especially in the wet.

Secondly, as the wheels land on the road, after hitting a bump and then flying through the air, they gradually make ruts. Not only can these ruts collect water, they can also make steering difficult. This water can soak through into the foundations and weaken them.

Thirdly, as a wheel rolls over a section of road, it squashes the road, which then expands once the wheel has passed. This process eventually leads to cracks in the road.

To compare the road damage caused by different vehicles, just look at a 40-tonne 10-wheel truck and a 1-tonne 4-wheel car. According to the road-planners, in a single pass the truck causes 160 000 times more damage than the car! And one type of damage is the corrugation.

Now there were many theories about how these corrugations formed, but it was Keith B. Mather from the University of Melbourne, way back in 1962, who did the

Death Penalty for Parking

King Sanherib of Assyria, around 700 BC, was so proud of his magnificent paved road that he brought in the death penalty for parking! His processional road, joining the main temples in his capital of Nineveh, was about three times wider than a modern 6-lane divided highway. King Sanherib decreed that anyone who parked their chariot on his road, or even erected a building that overshadowed his road, would be executed by the particularly messy method of being impaled on the pinnacles on the roof of their own house!

Cruise Control Can Kill Tyres

There seems to be a brand new disease of tyres that is directly caused by 'cruise control'. Cruise control lets you cruise on the highway at a fixed speed.

According to Bruce Reilly, the managing director of Truck Align at Narellan in NSW: 'When travelling at a stabilised speed the tyres actually induce an area of harmonic vibration. This never happened in the old days, because the speed always varied. But with a constant speed setting, once the harmonic vibration has started it can continue for hours. This can slowly destroy a tyre as it literally shakes itself to pieces.'

experiment that finally gave the real answer.

He started with a metal arm about 1 metre long that had a pivot at one end, and a small wheel at the other. He then drove it around in circles with a variable-speed electric motor. The wheel travelled on a circular concrete track, which he covered with different materials. He was very thorough – not only did he use a fine-sifted sand, but also different grades of coarse sands. In fact, he even ran the wheel over various dry products like gravel, rice grains, split peas and even sugar!

He found that provided these materials did not stick together, and were dry, he could always generate corrugations. The corrugations would come quickly if he had hard tyres, but would come slowly with soft tyres. They came readily in dry materials, but very slowly, if at all, if they were wet. It made no difference to the corrugations if the wheel was a driving wheel, or a driven (or rolling or idling) wheel. And he found that the major factor in generating corrugations was the road speed of the vehicle.

It was all because of bumps. You can never make a road perfectly smooth. There will always be tiny little bumps. On his set-up, once the wheel got up to a certain minimum speed (about 7 kph), it would bounce in a little hop after hitting one of these tiny bumps. When the wheel came down and hit the sand, it would spray sand both forwards and sideways off the track, leaving behind a little crater – which would then be the valley of a corrugation. As the wheel came up out of the valley, it would jump into the air again, and so the pattern of valley-and-mountain would repeat itself.

As they began to appear on the smooth road, the first few corrugations would be quite shallow, and very close to each other. But as the corrugations got deeper, they would gradually move apart from each other, until their height and their distance apart had settled into a stable pattern. Once this stable pattern of corrugations was set up, then the entire pattern of corrugation would migrate down the road in the direction of travel of the wheel. In the Australian outback, engineers have seen corrugations heading in opposite

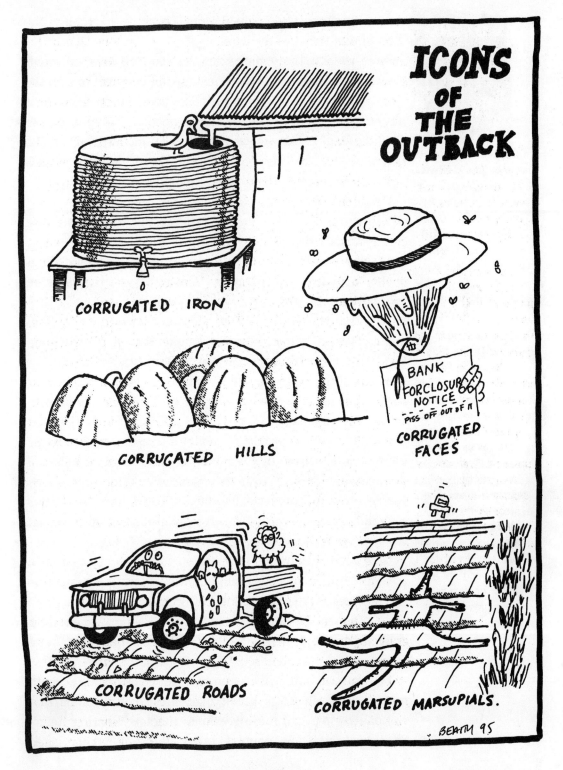

Corrugations Versus Engine

Corrugations can make the body of a car twist, but so can a powerful engine. The famous Ford Falcon GT HO was supposed to twist the body so much with its powerful engine (over 215 kW) that the blinker would sometimes switch on!

According to Joe Kenwright, writing in *Car Australia*: 'Wound up on a drag strip and thrashed mercilessly the powerful V8 will impose such huge torsional forces on the body during a slammed one-two gear change that the right indicator will come on as the steering column winds up from the twist. For serious HO punters, if you don't shock the right hand indicator on, you're not trying hard enough.'

directions on each side of the road – each set heading in the direction of travel of the cars!

Now even if the vehicles on the road all look different, they're all pretty similar as far as making corrugations on the road is concerned. In general, the vehicles on the open road all tend to travel around the same speed. So a bump on the road that makes the wheels of one car bounce, will also make the wheels of any other car bounce. These bouncing wheels will all tend to land around the same point. And that's how the corrugations form.

Of course, the faster the road traffic, the further apart are the corrugations. French engineers working in the desert roads of North Africa found that the corrugations were about 1 metre apart, while Mather found that corrugations in Australia averaged out to about 0.75 metres.

However, since this research was done, a whole new type of 4-wheel drive suspension has arrived on the scene – the long-travel, high-compliance coil suspension started by Range Rover, and then copied, with varying degrees of success, by the Japanese. If we could manage to get the vehicles with different suspensions to travel on different (but otherwise identical) roads, it would be really interesting to see if the corrugations caused by the long-travel, high-compliance coil springs were different from the corrugations caused by short-travel, low-compliance leaf springs. (You'd have to do this experiment after the road was freshly graded.) I suspect that the corrugations would be different.

Now corrugations don't appear only on dry outback roads. You can occasionally see them on bitumen and even concrete roads – usually after some kind of break in the road surface. They're especially easy to see in the wet, at night, when water collects in the troughs, and reflections make them more obvious.

You can see corrugations on railroad tracks, and the train makes so much noise going over these corrugations that railroad people call these sections of track 'Roaring Rails'. Skiers in the snow country often find rough corrugated washboard patterns on a well-travelled ski trail. If you have ever replaced wheel bearings because they were noisy, you'll see regular wear patterns on the roller or ball bearings.

The overhead copper conductors that trains and trams get their electric power from often show corrugations. The little valleys are the locations where the pantograph (the device that transfers the power from the wires to the train) broke contact from the copper wire, and set up an electric arc that vaporised away some of the copper wire.

Mather discovered that the important factor involved in making road corrugations is the high stress developed by the landing wheel, rather than the material that it sprays out. The stress makes the sand spray out. The stress also makes the bitumen flow plastically in paved roads, and the steel flow plastically in the balls of a ball bearing.

In his research, Keith Mather discovered another unfortunate result. *A smooth, flat road is unstable, and will rapidly turn into its stable form – the corrugated road.* He saw many cases where a smooth road turned into a corrugated road – but he never saw a corrugated road turn into a smooth road. The wheels never did remove material off the tops of the tiny mountains, and put it into the tiny valleys.

There aren't many things we can do to prevent corrugated roads in the outback. If we had the money, we could use something like concrete to make the road surface so tough that it's hard to deform – but that is too expensive for a country like Australia with so many roads and so few people. We could add silica gel (which absorbs water) to the road to keep it damp – but it's expensive, and has to be constantly replenished. It seems that corrugations on our outback roads will be around for a long time. So when you're driving, all you

How to Estimate Road Wear

To estimate road wear, road-planners use the famous 'fourth-power' rule of road wear. This 'fourth-power' rule is based on endurance tests that were carried out in the late 1950s by the American Association of State Highway Officials. It's not all that accurate and reliable, but it is a starting point.

According to this law, if you double the load on a wheel, you don't just increase the road damage by a factor of two, but by a factor of 2^4, which is sixteen! Compare a 40-tonne 10-wheel truck with a 1-tonne 4-wheel car.

Each of the 10 wheels of the truck carries 4 tonnes, and 10 multiplied by 4^4 is 2560. Each of the four wheels of the car carries $\frac{1}{4}$ tonne, and four multiplied by $\frac{1}{4}^4$ is 0.01525. The ratio of 2560 and 0.01525 is 163 840.

So if you believe the 'fourth-power' law, this is how a truck causes 160 000 times more damage *in a single pass* than the car!

can do is to try to go outside the resonance speed of the corrugations – either faster or slower.

But corrugations have at least one use. Place your dirty clothes, together with water and washing powder, into a bucket with a screw-cap lid. Put the bucket in your car, and after a few hours of driving on a corrugated road, your clothes will be clean!

References

Scientific American, January 1963, 'Why do roads corrugate?' by Keith B. Mather, pp.128–136.

Facts & Fallacies, Reader's Digest (Australia) Pty Ltd, 1989, 'A long, long trail a-winding', pp.122–123.

Scientific American, November 1989, 'The world's oldest road' by John M. Coles, pp.78–84.

New Scientist, No. 1747, 15 December 1990, 'Roads to ruin' by David Newland, pp.33–40.

Car Australia, August 1994, 'Ho, Ho, Ho' by Joe Kenwright, pp.58–65.

Daily Telegraph Mirror (Sydney), 17 June 1995, 'Airbags on a roll', p.121.

BEAUTY

OUR SOCIETY THINKS 'BEAUTY' is very important. For example, one of the few professions in which women usually earn more than men is modelling.

The problem of what beauty is exactly has been worrying thinkers for thousands of years. In fact, a whole branch of philosophy, called aesthetics, was invented to deal with these difficult concepts of beauty and ugliness. The philosophers have had the field to themselves for a long time, but now the zoologists, botanists and even the neuroscientists are talking about 'beauty'.

Maybe *Roget's Thesaurus* got closer than it realised when it gave one meaning for 'beauty' as 'symmetry'. Symmetry means being the same, or even, on each side. Now most of us humans (the right-handed ones, at least) have our right foot slightly bigger than our left foot. In other words, our feet are *non-symmetrical*, or *asymmetrical*. Photographers have long known that there's always a slight difference between the left side of your face and the right side of your face. And so have many actors, who, since the early days of Hollywood, have preferred to be photographed on only one side of their face – their best side, of course.

Over the last few years, though, the biologists have looked at the animal kingdom, and they've made a few discoveries about how symmetry relates to fitness and beauty.

Firstly, symmetry is definitely related to **fitness**. Horses that are more symmetrical can run faster than horses that are less symmetrical. In one study, biologists measured some 10

Symmetry in Plants and Insects

Even bees love symmetry. It turns out that within a species, symmetrical flowers produce more nectar than asymmetrical flowers. In one study, bees were twice as likely to visit a symmetrical flower as compared to an asymmetrical flower.

When a researcher turned a symmetrical flower into an asymmetrical flower by removing a part of some petals, even though it had just as much nectar as before, it suddenly became unattractive to bees!

features on 73 thoroughbreds – features such as the thickness of the knee, or the width of the nostrils. As the differences they could measure were quite small, these probably had no direct bearing on how fast the horse could run. The symmetry is probably a clue to the horse's hidden increased fitness.

In fact, symmetry is probably a good indicator of general underlying health, vigour and strength.

Our imperfect world is full of stress – like nasty chemicals (to poison us), germs (that want to live off us) and bad climate (to wear us out). It seems that only those individuals who are lucky enough to inherit a sturdy genetic makeup, and are also lucky enough to get good nutrition and a healthy environment while they're growing, will end up being quite symmetrical. Certainly, if you expose baby fruit flies to nasty chemicals, or parasites, or very hot or cold temperatures, they will grow up with differences between each side of their body.

In another study, one group of Israeli scientists claims that women who suffer from infectious diseases during their pregnancy are more likely to give birth to asymmetrical babies. They also claim that these asymmetrical babies will grow up to suffer more heart disease than their more symmetrical cousins.

Secondly, the biologists found that differences that do exist in animals between one side of the body and the other are much greater in the so-called **secondary sexual characteristics** – antlers or horns, breasts or tails, or even enlarged canine teeth or spurs.

The scientists who study elk noticed that the males that had the largest harems of females had the most symmetrical antlers. The usual way that the antlers on a male become asymmetrical is by losing a segment or two in a fight with another male elk. Of course, a number of elk would have the genetic background to give them greater symmetry and others would have better nutrition, but most male elk would enter adult life with some degree of asymmetry. If an elk was very powerful and strong, and a very good fighter, though, he would be less likely to lose fights, and would probably have a more symmetrical rack of antlers.

Thirdly, animals that are more symmetrical are more likely to **attract** a mate.

Scientists have found this preference for symmetry in female zebra finches. The biologists put some circular coloured bands on the legs of male zebra finches. The females preferred those males that had the same number of bands on each leg. Other biologists found that female Japanese scorpion flies much prefer a symmetrical male scorpion fly as a mate than an asymmetrical male. Another scientist found that he could turn attractive (in terms of getting a mate) male swallows into unattractive male swallows (and also ruin their chances of a good sex life) by clipping their tail feathers asymmetrically with scissors!

But if we have a vague idea of what other members of the **animal** kingdom find attractive, we're on much shakier ground with **humans**. Just like in the other animals, symmetry in humans is probably a part of what we think is beautiful. Of course, it's very complex, and there are many other factors involved, but symmetry seems to be one of the big ones.

Various studies have shown that if a man's body is very symmetrical, he will begin his sex life earlier, he'll have more sexual partners, and he'll be a better lover!

The first study showed that the more symmetrical a man's body is, the earlier he begins an active **sex life**.

The second study, of 122 university students, showed that the students who were more symmetrical in their faces and bodies had **more sexual partners** than their unbalanced fellow students.

The third study claimed that symmetrical men make **better lovers**. This study involved symmetrical and lop-sided men. (One major problem in this kind of study is in

Symmetry and Secondary Sexual Characteristics

In other members of the animal kingdom, the secondary sexual characteristics show the greatest degree of asymmetry. One of the secondary sexual characteristics for women is the development of breasts. Women's breasts can differ in circumference, from one side to the other, by as much as 30 per cent!

One study of 50 American women claimed to show a relationship between the number of children that a woman had and the symmetry of her breasts. According to this study, women with more evenly sized breasts had lots of children, while women with differently sized breasts had fewer children.

BEAUTY AND SYMMETRY IN THE SOUTH PACIFIC - POST MURUROA

1. ASYMMETRICAL

NOT
VERY
BEAUTIFUL

2. SYMMETRICAL

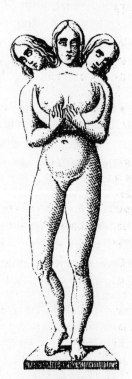

TOTALLY
GORGEOUS

BEATTY '95

The Philosophy of Beauty

Practically everybody who has tried to wrestle with the concept of beauty has come up with a concept that is very different from other people's concepts.

Plato, the Greek philosopher, claimed that all of the objects that existed in our world were just flawed and imperfect examples of the ideal, or timeless, perfect form. So somewhere, there was a theoretically perfect spear, or a theoretically perfect drink bottle. And the closer that any physical object approached this unachievable perfect form, the more beautiful it was.

Aristotle, his fellow countryman, took a different approach, however. Some three and a half centuries before the birth of Christ, Aristotle said that: 'Beauty depends on size as well as symmetry. No very small animal can be beautiful, for looking at it takes so small a portion of time that the impression of it will be confused. Nor can any very large one, for a whole view of it cannot be had at once, and so there will be no unity and completeness.'

In 1753, the German philosopher Alexander Gottlieb Baumgarten introduced the word 'aesthetics', which came from the Greek word 'to perceive'. Another German philosopher, Immanuel Kant, claimed that objects were thought to be beautiful when they were beyond personal needs or interests. Yet another German philosopher, Hegel, thought that beauty was everything that our human spirit found uplifting and pleasant.

The French poet Charles Baudelaire thought that 'all forms of beauty...contain an element of the eternal, and an element of the transitory'. Jean Luc-Godard, the French filmmaker, agreed when he said that 'beauty is composed of an eternal invariable element, whose quantity is extremely difficult to determine, and a relative element, which might be, either by turns or all at once, period, fashion, moral, passion'. The French author Stendahl had a different point of view. He thought that 'beauty is only the promise of happiness'.

In 1884, the playwright Oscar Wilde said in *The Value of Art in Modern Life:* 'I have found that all ugly things are made by those who strive to make something beautiful, and that all beautiful things are made by those who strive to make something useful.'

And Jean Kerr, an American playwright, wrote in 1958 in *The Snake has All the Lines, Mirror Mirror on the Wall:* 'I'm tired about all this nonsense of beauty being only skin deep. Isn't that enough? What do you want— an adorable pancreas?'

Symmetry — Fitness or Not?

Some research claims that in a sexually mature animal, symmetry is some sort of indicator of general fitness. But another series of experiments has a different explanation for the importance of symmetry!

Swedish zoologists set up artificial networks of nerves that had the ability to learn — so-called 'neural networks'. To their surprise, they found that it was easier for these neural networks to recognise symmetrical shapes. Maybe our nervous systems are just better at recognising symmetrical objects when we happen to look at these objects from different angles. This could be why nature seems to love symmetry in some animals and plants.

setting up the control group — for example, how can you randomly allocate sexual partners, and still get approval from the ethics committee?) It found that during sexual intercourse, women who made love with the most symmetrical men had orgasms 75 per cent of the time. But women who made love to the most asymmetrical men had orgasms only 30 per cent of the time. And the more symmetrical man was more likely to have his ejaculation at the same time his female partner was having her orgasm! In fact, another finding according to this study, was that symmetry also meant more babies.

Now it's a bit strange to imagine that a few grams of flesh on one side of the body rather than the other side should make any difference to one's fertility, fitness or attractiveness. But maybe, just like in the animal kingdom, symmetry in humans could be a measure of general underlying fitness and vigour.

The American author Naomi Wolf claims in her book *The Beauty Myth* that 'beauty' does not exist objectively and universally. She claims that your idea of beauty depends on how, where and when you were brought up. She agrees with the old saying 'Beauty is in the eye of the beholder'.

But one study by a team of psychologists, Perrett, May and Yoshikawa, disagreed with this. In fact, it came up with a few surprises. The psychologists superimposed photographs of hundreds of women, and showed these averaged (or blended photos) to their test groups. The photos were of Caucasian and Asian women, and the test groups were made up of Caucasian and Asian men and women. The psychologists found two main results.

The first finding was that the test group did not choose the so-called 'average' face as the most attractive face. This was a bit of a surprise, because some previous studies had claimed that the most attractive face was the 'average' face. In this study, the most attractive faces were quite different from the photographically averaged faces.

And secondly, they found that the female face chosen as more attractive had higher cheek bones, larger eyes and a thinner jaw. They also found that if a female face was small around the mouth and the chin, it was considered more attractive. So a small distance between the mouth and the nose, and between the mouth and the chin was considered most attractive. It turns out that as women grow older, the lower part of their face begins to fill out. Maybe *younger* equals *more fertile* equals *more attractive*. (This study judged only female attractiveness. We're waiting for the follow-up study to see if there is a link between male symmetry, sperm count and male attractiveness.)

But all the studies done so far have involved small numbers of people – usually less than 150, and mostly, around thirty. And most of these studies have not yet been repeated by other scientists, to check their accuracy.

It seems that as far as we humans (and science) are concerned, beauty may not entirely be in the 'eye of the beholder'.

Perhaps we should let the great wordsmith John Keats have the last word:

> *'Beauty is truth, truth beauty'– that is all*
> *Ye know on earth, and all ye need to know.*

Symmetry and Disease

People with asymmetrical teeth tend to have more micro-organisms in their mouths.

People suffering from schizophrenia tend to have their fingerprints more unmatched between their left and right hands than non-schizophrenics do.

References

The Oxford Companion to the Mind by Richard L. Gregory, Oxford University Press, 1987, pp.8–10.

Nature, Vol. 367, 13 January 1994, 'Preference for symmetric males by female zebra finches' by John P. Swaddle and Innes C. Cuthill, pp.165–166.

Nature, Vol. 368, 17 March 1994, 'Beauty and the beholder' by Nancy L. Etcoff, pp.186–187.

Nature, Vol. 368, 17 March 1994, 'Facial shape and judgements of female attractiveness' by D.I. Perrett, K.A. May and S. Yoshikawa, pp.239–242.

Nature, Vol. 372, 10 November 1994, 'Symmetry without fear' by Mark Kirkpatrick and Gil G. Rosenthal, pp.134–135.

Nature, Vol. 372, 10 March 1994, 'Symmetry, beauty and evolution' by Magnus Enquist and Anthony Arak, pp.169–172.

Discover, November 1994, 'Symmetry: boring but fast', p.20.

FROG AND MOUTH ANTIBIOTICS

Antibiotics and Mouthwash

Mouthwashes seem to be pretty useless at killing bacteria. In one study, mouthwashes were added to saliva, which has millions of bacteria. The saliva samples were exposed to different mouthwashes for 24 hours. Four of the mouthwashes had a little bit of activity against the bacteria, some had very little activity, and one had virtually none at all.

Think about it. The best of the mouthwashes kills only a few bacteria, even after being exposed to the bacteria for a whole 24 hours. You can see how little effect mouthwashes must have when they are merely swished through the mouth for a minute or two!

EVER SINCE WE humans started using antibiotics, bacteria have fought back, by becoming resistant to our drugs. Every time we came up with new drugs, the bacteria evolved resistance to them. For a while, it looked as though we were running out of new antibiotics, and the germs were going to win. But luckily, the frogs have jumped in to give us a hand.

Medical people, or healers, have known for thousands of years that some substances or chemicals made by one living creature (and not harming it) can kill other living creatures.

Two and a half thousand years ago, the Chinese wrote that the mould that grows on soybean curd could cure boils. But this was known to the healers of Egypt and Mesopotamia even earlier. The mould was making a chemical (streptomycin) that killed the bacteria that caused the boils. In fact, even today, this mould gives us the streptomycin that is our main defence against the bacterium that causes the plague – *Yersinia pestis*.

In the late 19th century, various scientists reported how one micro-organism could interfere with another micro-organism.

In 1874, William Roberts of Manchester saw that fungi would interfere with the growth of bacteria, and vice versa. Presumably, the fungi were making and releasing chemicals that acted against bacteria (antibiotics), while the bacteria were making and releasing chemicals that acted against fungi (anti-fungals). He even noted that one fungus, *Penicillium glaucum*, was never attacked by bacteria.

In 1877, on the other side of the English Channel, Louis Pasteur was doing experiments with the bacterium that caused anthrax. He could easily grow the anthrax bacterium in a nutrient solution – except when this solution was accidentally contaminated by fungi. He and his co-worker, Jules Francois Joubert, suggested that this fact could one day be used to stop infections, but they did not follow this line of research.

Around 1900, the German bacteriologist Rudolf von Emmerich purified a chemical called pyocyanase. This chemical easily killed the bacteria which cause diphtheria and cholera – but only in the test tube on the laboratory bench. It didn't kill these bacteria in humans.

In 1928, Alexander Fleming was trying to grow a bacterium called *Staphylococcus aureus* – the now-famous Golden Staph. Usually, he could grow it quite easily, but every now and then, a fungus would contaminate his growing plates, and the bacterium would not grow where the fungus was. The fungus was called *Penicillium notatum*, and he isolated a chemical from it, which he called penicillin. Penicillin would stop the growth of *Staphylococcus aureus*. Unfortunately, he could get only very small and very impure quantities of penicillin. As a result, it didn't seem to be very effective in humans as a drug to kill bacteria.

The modern era of antibiotics began in 1936 with the drug sulphanilamide. The German chemist Gerhard Domagk found that this drug, based on a synthetic dye called prontosil, would stop many different types of bacteria from multiplying. It would not kill them, but by stopping them from increasing their numbers, it gave the body's immune system time to regroup, and kill the invading bacteria.

In 1939, Rene Dubos at the Rockefeller Institute successfully followed up the observation of Pasteur and Joubert. Dubos believed that one micro-organism could block the growth of another micro-organism. He could see that in

African Frog and Pregnant Australian Women

In the days before cheap and reliable pregnancy tests, the African clawed frog, *Xenopus laevis*, was used to diagnose pregnancy in Australian women!

Urine was collected from a woman who wanted to know if she was pregnant. This urine was injected into *Xenopus laevis*. If the woman were pregnant, she would be making special hormones of pregnancy. These hormones would make *Xenopus laevis* produce eggs.

So the first hint of the existence of many Australians was that in a laboratory somewhere, *Xenopus laevis* began making eggs!

Why do Frogs Croak?

Frogs croak to attract a female. The trouble is that they also attract unwelcome prowlers. In a typical pond with 450 frogs assembled in a night chorus, 12 frogs were eaten every hour by bats, while opossums, bull-frogs and even underwater crabs mopped up a few more. But there's safety in numbers – the more frogs in the chorus, the safer each individual frog is.

In the Panama Canal there is a tropical forest island called Barro Colorado. A frog called the túngara is noted for being very loud – as loud as a heavy truck at 15 metres, or 90 decibels! One evening a male was counted as croaking 7244 times. The trouble is that croaking males burn up energy twice as fast as the silent males – but at least they get to party! In some cases, a frog who is continuously croaking *and* fighting, can lose up to one-third of his body weight. But it pays off. Female frogs seem to be attracted more to an energetic male – or at least, to one who can put more sound energy into his song.

ordinary soil there was a balance between micro-organisms. No one single micro-organism ever gained the upper hand and wiped out all the other micro-organisms. After much hard work, he isolated from a bacterium living in ordinary soil an antibiotic called tyrothricin. It worked just fine on skin wounds, but it was much too toxic to be swallowed.

Also in 1939, a group of scientists at Oxford University, Howard Walter Florey, Ernst Boris Chain and Edward Penley Abraham, began attempts to purify penicillin. The 'Golden Age' of antibiotics began in 1941 when they succeeded in getting pure quantities of penicillin. There were still major difficulties in mass producing penicillin, however. But by using some of the technology that is used to brew beer, they did achieve mass production. The first batches of penicillin became available in 1943, and were used by the military.

Penicillin was truly a miracle drug when it was first introduced. It worked quickly and effectively against pneumonia, meningitis and hundreds of other deadly diseases. However, after a while, bacteria became resistant to some of the penicillins, so the pharmaceutical companies came out with new types of antibiotic drugs. But then the bacteria mutated again. After many such cycles, there are now strains of gonorrhoea and tuberculosis that have mutated so much that they are resistant to practically all known antibiotics.

While a generation for humans takes roughly 25 years, a generation for bacteria takes only a few days. So the bacteria can very rapidly evolve resistance against our antibiotics. This means that we're stuck in a continuous cycle of always needing new antibiotics. (Of course, if we don't use antibiotics for various infections, the bacteria won't evolve –

but millions would die.)

Today, about 30 per cent of all patients who enter hospital will get at least one course of antibiotics. Antibiotics have saved millions of human and animal lives.

So far, most antibiotics have come from **micro-organisms**, such as fungi and bacteria. But recently, pharmacologists have been looking at antibiotics from **animals**, such as frogs.

For thousands of years, ancient healers have used parts of frogs and newts and other animals (as well as herbs) in their brews. In 1986, Dr Michael Zasloff, chief of Human Genetics at the US National Institute of Health, noticed one of his frogs swimming through a tank filled with dirty stagnant water. As part of his research, he was studying *Xenopus laevis*, the African clawed frog. It's a very sturdy, resilient and hearty frog, and so it's used in laboratories all over the world. He had just done an operation on the frog in question, so it had a fresh scar. But even though the water was filthy, the frog was healing perfectly. He suddenly realised that over the many years that he had worked with this species of frog, he had never seen an infected wound on a frog!

Zasloff spent months examining the skin of the frog, and he found a brand new family of chemicals there. These chemicals can kill bacteria, fungi and parasites. He calls the chemicals 'magainins', from the Hebrew word for 'shield'.

When a shark expert told Zasloff that pregnant sharks would flush their bodies by drinking sea water that was lousy with bacteria, he suddenly realised that sharks too must have their own natural antibiotic. He went looking and he found yet another type of antibiotic – a steroid.

Difference Between Bacteria and Virus

Bacteria and viruses are quite different, and drugs that kill bacteria usually won't kill viruses.

A bacterium is a living creature. It has a cell wall, and it tries to protect the integrity of what lies inside that cell wall. A bacterium will eat food and will excrete waste products. We humans have many antibiotics, and these can kill bacteria. Many of these antibiotics will interfere with some part of the metabolism of a bacterium.

A virus is almost a living creature, but not quite – It doesn't have its own metabolism. It relies on getting inside a cell (human, animal, fungus, plant, bacteria etc) and taking over the DNA of the host. The host will then make many copies of the virus, and these copies will then leave the host, to infect other cells. Sometimes a virus will harm a cell greatly, and sometimes it won't harm it at all. We have only a handful of drugs that will work against viruses.

The easiest way to appreciate the difference between a bacterium and a virus is to think of a giant ocean full of food (fats, sugars, proteins, salts, minerals etc). Our hypothetical ocean is full of food, but there is not a single living creature in it.

If you add one single bacterium to the ocean of food, it will multiply like crazy. Very shortly, the ocean will be full of bacteria.

If you add one single virus to the ocean of food, it will not multiply. It can multiply only inside the cells of a living creature.

Antibiotics in Ants and Cockroaches

Professor Andrew Beattie from Macquarie University has discovered antibiotics in bull ants. They are in secretions from glands on the ant's chest, and will kill bacteria and yeast.

Professor Richard Karp, Professor of Biology at Cincinnati University in Ohio, has found a strange new antibiotic in cockroaches. He was first made curious about cockroaches because of their long lifespan. Most insects live for just a few weeks, or perhaps a season, but cockroaches can live for four years.

If they survive that long, he reasoned, they must have a good immune system. He found three different types of immune system response – the first against bacteria, the second against soluble toxins, and the third against foreign transplanted tissue. But these responses are different from similar human ones.

In our human response against bacteria, our antibodies latch onto the bacteria. Then other parts of the immune system come and attack the 'flagged' bacteria.

In the cockroach, however, the immune system chemicals actually find and then kill the bacteria. These anti-bacterial chemicals are very big, and would get broken down by our digestive enzymes. But Professor Karp thinks he can isolate the active sections of the chemicals, and then give them as a medicine.

The antibiotics from the frogs and the dogfish sharks have already been patented. However, there's a lot that can go wrong between discovering an antibiotic and releasing it on the market. Some antibiotics will kill bacteria in a glass tube on the laboratory bench, but they won't kill the same bacteria in a living animal. Other antibiotics turn out to have savage side effects when they are used at concentrations high enough to kill the bacteria. But the frog antibiotic has already passed the test on laboratory animals, and is now being tested on humans. It could well be on the market in 1996 to treat diabetic foot ulcers and impetigo (a nasty skin infection).

Other chemicals besides antibiotics have been found in frogs. One frog from Ecuador is loaded with a previously undiscovered painkiller, epibatidine, that is 200 times more powerful than morphine! And three species of Australian tree frogs exude a chemical which kills not only bacteria, but even viruses, which are usually much harder to kill.

It seems that most animals have some sort of chemical defence. Some animals (like the cane toad) are loaded with nasty chemicals which keep *big* predators away, while other animals make antibiotics which protect them against infection by *tiny* predators like bacteria.

Back in 1993, some people thought that mammals might make their own natural antibiotics too but at that time, no scientist had gone looking for them. But in 1995, one scientist found a brand new antibiotic in the mouths of cattle.

The mouth is a very busy place. Over the 70 or more years of your life, you'll use your mouth to eat about 40 tonnes of food. All sorts of germs will hitch a ride into your

Frogs and Stomach Ulcers or Birth Through the Mouth

Most frogs are careless and indifferent parents. They simply abandon their offspring in a convenient pool of water – such as a pond, or even a shallow puddle in the middle of the road. But some frogs do care for their young. They actually carry their offspring around until they are big enough to survive in the open world. The male of a tiny Queensland rainforest frog *Assa darlingtoni*, has tiny hip pouches. The growing tadpoles wriggle into these hip pockets, and then stay there until they are big enough to survive. But as they get bigger, they take up more room, and they end up squashing daddy's stomach so much that he cannot eat at all. He's a 'glutton for punishment'!

One frog even goes one step further. This frog is called *Rheobatrachus silus* – *rheo* means 'stream' or 'flow', *batrachus* means 'frog', and *silus* means 'pugnosed' (this frog does have a squashed-up or blunt snout). This very small (5 cm long) dull-brown frog lives under rocks in streams. The mother swallows the fertilised eggs or tadpoles into her stomach, which turns into a uterus or womb. After eight weeks, she gives birth to about 25 young frogs through her mouth.

When she swallows her babies, they release a chemical which stops her stomach from producing hydrochloric acid. Hydrochloric acid is used to help digest and break down the food in the stomach. This acid is good for food, but not for growing babies. As the young get bigger, Mother frog doesn't eat at all. She has to live entirely off her stored fat. By the time the babies are ready to leave, she's looking pretty scrawny. The babies grow bigger, and so her stomach has to get bigger. Her stomach wall gets thinner. Eventually the babies weigh about 40 per cent of her total body weight. By this time, her lungs have been so squashed by her babies, they have collapsed! So she has to breathe through her skin, and through a special mucous membrane on the roof of her mouth.

Then, over a period of five days, she gives birth to the babies. She relaxes the muscle at the top of the stomach, has a tiny vomit and opens her mouth, and as if by magic, one or two babies appear in her mouth. If the babies want to come out, they hop away. But if they don't, she waits for a reasonable time with her mouth open, and then she swallows them again, and they go back into her stomach. But sometimes, when she is under stress, or simply going to starve if she doesn't give birth soon, she just rises to the surface, opens her mouth, contracts her muscles and squirts her babies out, like tiny rockets, over a distance of a metre! Then her stomach gradually returns to normal.

The scientists did an experiment. They waited until a mother frog had given birth to her babies. Then they gave her a meal of 10 very delicious, appetising, enticing, delectable and tasty mealworms, to tempt her ➤

▷ to eat. After four days, she began to eat. When they examined her stomach, it had returned to normal.

But what about that special chemical given out by the tadpoles that stops the production of stomach acid? It could be a great anti-ulcer drug. In fact, the drug which is the number one seller in the whole world is an anti-ulcer drug. So this substance could be used to earn squillions of foreign cash for Australia. Unfortunately, the frogs have not been seen since 1986!

How Antibiotics Work

Antibiotics can work against bacteria in several ways. They can interfere with the manufacture of the cell walls of the bacteria, or they can make the cell wall of the bacteria leaky. They can affect the metabolic machinery of the bacteria so that it doesn't make proteins at all, or so that it will make only badly misshapen proteins. Some antibiotics interfere with the operation of the bacteria's DNA, while others block essential metabolic pathways that the bacteria need for their day-to-day living.

mouth along with that food – germs such as bacteria, viruses and fungi. For a long time, though, nobody asked the obvious question – why, with all these germs coming in, do you hardly ever get infections of the mouth?

(Actually, the mouth can *sometimes* be easily infected. This usually happens only when your immune system is suppressed – for example, if you have AIDS, leukaemia, or are having chemotherapy for cancer. In these cases, you can have lots of warts, caused by the papilloma virus, or lots of ulcers from the herpes virus.)

There's an old saying in science: 'It's not the answer that gets you the Nobel Prize, it's the question!'. Once you have asked the right question, many likely answers become obvious. In this case, one possible answer is that something in your mouth might make antibiotics.

There's always been a very subtle hint that the tongue might be a good place to look for new antibiotics. When animals or children get a wound, they will often lick that wound. But nobody ever followed up on that hint.

Your tongue is a strong muscular organ that you use for swallowing, for speaking and for tasting. Muscles are a bit like octopus straps – all they can do is pull. But the muscles in the tongue are so cleverly arranged that the tongue is the only muscle in the entire body that can *push*!

Each day, we swallow about 2500 times, and once again, the tongue helps us do that. In fact, swallowing is so essential that our brain has been hard-wired to make our tongues do sucking motions even before we are born. The tongue is also important for speaking, and the tongue and

Penicillin and the Very Good Experiment

One of the very first patients to use penicillin helped set up a scientifically excellent, but personally distressing, experiment. The type of experiment involved is called a longitudinal experiment, where you can follow the subject through time.

In 1941, during World War II, an English police officer was suffering a massive infection of his bone (called osteomyelitis). He was given the world's whole supply of penicillin, and he got better. Once all the penicillin was gone, though, he got worse. The medical scientists then managed to purify some penicillin from his urine! They gave it to him, and he got better again. Finally, they ran out altogether, and he died.

For the police officer it was very distressing. As an experiment, it was excellent (even if the sample size was rather small).

the teeth have been cleverly wired-up in the brain so that no matter how fast we speak, we hardly ever bite our tongue with our teeth.

The tongue is one of the few organs in the body that does not always get weaker with age. It can actually get stronger! There are probably two reasons for this. One is that the tongue has a very good blood supply – about 10 times greater than the blood supply to the other muscles of the body. The other reason is that you are always using your tongue, so it doesn't get a chance to get lazy.

Because we're running out of antibiotics, the drug companies are desperately looking for new ones. They're sending researchers to look in sewers and swamps, in rainforests and ocean outfalls. And the latest place that they have looked is inside the mouth – and with a ready source of cows, and their tongues, at hand, cows were as good a place as any to begin looking.

That particular area was explored by Barry S. Schonwetter and his colleagues at the Magainin Research Institute in Pennsylvania. They were curious about why the mouth hardly ever gets infected, even though it is exposed to many different types of germs. So they had a close look at the upper surface of the cow tongue, both on damaged and undamaged areas.

They were able to extract a whole bunch of chemicals that would stop both bacteria and fungi. They looked very closely at the most abundant chemical, which they called Lingual Antimicrobial Peptide or LAP. They found that LAP would stop *E. coli*, *Pseudomonas*, *Staphylococcus aureus* (Golden Staph), and *Candida albicans*, the infamous yeast.

They also found that LAP was concentrated around the cuts in the cow's tongue. They didn't find it in the tongue of an unborn cow, but they did find it in four-month-old calves. They're not sure whether LAP appears after the tongue gets first infected by germs, or whether it's genetically programmed to appear at a certain age. They have found LAP in other parts of the cow such as the conjunctiva of the eye, the trachea, the uterus, the bladder – and in fact, anywhere

Penicillin and the Guinea Pig

It was very lucky that penicillin was developed during World War II, when everything was done very rapidly. If they had tested penicillin on guinea pigs, they might have dropped it. You see, while penicillin is virtually harmless to us humans, it kills guinea pigs!

that there is a barrier between the outside world and the cow's body.

The drug companies have been looking for new drugs all over the world, but all the time, there was a whole bunch of undiscovered drugs right under their noses!

References

New Scientist, No. 1823, 20 May 1992, 'Potent painkiller from poisonous frog' by John Emsley, p.14.

Popular Science, August 1993, 'The frog treatment' by Robert Langreth, pp.58–59, 78.

Science, Vol. 267, 17 March 1995, 'Microbes get licked' p.1573.

Science, Vol. 267, 17 March 1995, 'Epithelial antibiotics induced at sites of inflammation' by Barry S. Schonwetter et al, pp.1645–1648.

New Scientist, No. 1990, 25 March 1995, 'Nature's own antiseptic makes mouthwash redundant' by Peter Aldhous, p.16.

NON-LETHAL WEAPONS

NEARLY 2500 YEARS AGO, the famous Chinese martial strategist Sun-tzu wrote in his book *The Art of War* that even though you wanted to wage war, this didn't necessarily mean you had to kill people. It took a long time, but finally the military and the police are taking his advice.

The traditional practice in war has been to invade and break things and kill humans. But now some military strategists are looking at ways to disable, disorient and stupefy the enemy, as well as immobilise their weaponry. So there is a recent upsurge in interest in 'non-lethal' weapons, to give military leaders in the field a few extra options. In 1972, according to a US National Science Foundation report, there were over 30 non-lethal concepts listed. In 1995, the US Defence Budget put aside $US41 million (out of a total US Defence Budget of $US262 billion) to fund research into several 'interesting' concepts in non-lethal devices.

Now, of course, 'non-lethal' means different things to different people. According to the rather aggressive Edward Teller, famous for his work on the hydrogen bomb, small nuclear weapons count as non-lethal weapons! He has a vision involving small bombs about 130 times weaker than the Hiroshima atom bomb – in other words, equal to only 100 tonnes of TNT. He sees thousands of these small bombs being dropped accurately across the enemies' countries to destroy their infrastructure – bridges, railways, power stations etc. He even suggests dropping these mini-nukes on an 'unfriendly' country, but not exploding them until this

Arms vs People

Martin Luther, who led the Protestant Reformation, hated weapons. He said in 1569: 'Cannons and fire-arms are cruel and damnable machines; I believe them to have been the direct suggestion of the Devil. If Adam had seen in a vision the horrible instruments his children were to invent, he would have died of grief.'

On the other hand, Niccolò Machiavelli, the Italian philosopher and statesman, thought that weapons were essential to the smooth running of a state. In his famous book, *The Prince*, published in 1514, he says: 'The main foundations of every state, new states as well as ancient or composite ones, are good laws and good arms...you cannot have good laws without good arms, and where there are good arms, good laws inevitably follow...For among other evils caused by being disarmed, it renders you contemptible; which is one of those disgraceful things which a prince must guard against.'

But Mao Zedong, who helped found the People's Republic of China, and who led it for many years, realised that the will of the people is also important. As a guerrilla, he 'swam in the sea of the people'. In 1938, he wrote, in an essay called 'On Protracted War': 'Weapons are an important factor in war, but not the decisive factor; it is people, not things, that are decisive. The contest of strength is not only a contest of military and economic power, but also a contest of human power and morale. Military and economic power is necessarily wielded by people.'

country does something that offends you.

When the Cold War ended, the USA was left as the only military superpower. President Clinton of the USA has inherited a world without any almost-equal opponents. There has been a gradual trend away from wars *between* countries, to internal wars *inside* a country. In fact, in 1993, all of the 34 wars that raged on our planet were internal or domestic conflicts. So a new role for the United Nations seems to be peace-keeping – not to kill enemy soldiers, but to keep groups of people that hate each other away from each other. And that's why there has been such an upsurge of interest in *non-lethal* weapons.

But on a much smaller scale, the American Department of Justice wants some non-lethal weapons too. Janet Reno, the Attorney General of the USA, was very embarrassed by the debacle in April 1993 when the FBI tried ramming the walls of the stronghold in Waco, Texas, where the Branch Davidians were holed up. Eighty-six men, women and children died in the resulting fire. Surely, she reasoned, there must be a better way for a country as powerful and technologically competent as the USA to be able to disarm a small number of moderately armed civilians – without killing them! In April 1994, Janet Reno and John Deutch, the Deputy Secretary of Defence, signed a memo of understanding which said that the US military will allow certain non-military forces (FBI, police departments etc) to use certain non-lethal devices.

Non-lethal weapons seem to fall into two main categories – the ones used against **people**, and the ones used against **enemy weaponry**. Of course, there is some overlap and some weapons can be used against both.

One of the weird **anti-personnel** non-lethal weapons that the Pentagon is currently working on is a *sticky foam* that is a strange marriage between chewing gum and shaving cream. The foam supposedly immobilises people, without harming them. But if the foam covered their mouths and suffocated them, then it would be a lethal weapon. There are already functioning prototypes of this foam.

It was originally invented to stop terrorists from stealing nuclear weapons – it was to be sprayed all over the terrorists and the weapons. The Access Delay Technology Department of the Sandia National Laboratory at Albuquerque in New Mexico (which specialises in ways of protecting nuclear weapons) is now trying to bring this sticky foam out of its laboratory and into the street. It starts as a brownish liquid, and expands in volume some 50 times! Research is following two main paths – a thinner, more liquid version that will be used to inundate a room, and a thicker, more viscous version that can be sprayed onto a person some 15 metres away. This foam can be removed, but very slowly. Baby oil can remove the foam – but it would take about 45 hours to remove the foam from every square centimetre of the average person with a surface area of 1.73 square metres!

Another non-lethal weapon could be *very low frequency sounds* that, depending on the loudness, make people either vaguely unwell or queasy, or disoriented and dizzy, right up to bowel spasms, outright vomiting and uncontrollable diarrhoea. This could be generated by bursts of compressed air passed through giant horns. This project is still in an early stage.

Air bags are planned for the back of taxis and police cars to immobilise potentially dangerous passengers. These air

John Alexander

John Alexander, an ex-US Army colonel in charge of the Non-lethal Defence Program at the Los Alamos National Laboratory, is one of the leading proponents of non-lethal weapons.

He has also been interested in psychic ways of doing damage to the enemy, or of finding out their intentions. In 1980, he suggested in an article in *Military Review* called 'The Mental Battlefield', that a viable military option might be for soldiers to 'project their consciousness' into enemy bunkers to make the enemy sick, or to 'look' over their shoulders at their battle plans.

Alexander has a doctorate in education from Walden University in Florida, majoring in thanatology (the study of death), which he studied with the famous psychiatrist Elisabeth Kübler-Ross. In 1993, he organised a parapsychology conference on 'Treatment and Research of Experienced Anomalous Trauma', during which people discussed their own personal experiences with ritual abuse and abductions by aliens. In January 1994, he was hailed as an 'Aerospace Laureate' by *Aviation Week and Space Technology*.

Weapons Spending ①
or Nine Hospitals per Day

In 1987, the combined worldwide military spending was about $US1 trillion. Since then, it has dropped a little in each successive year.

This is such a huge number, that it is almost impossible to understand how big it really is – but one way to make sense of it is to realise that it can pay for nine new hospitals or universities per day! A trillion is a thousand billion. That means that the total military spending for 1987 was about $US1000 billion – which works out to $US2.7 billion each day.

Now consider a major teaching hospital – everything from operating theatres to bandages to intensive care units and X-ray machines. To set up the whole box and dice, including building it from scratch, and paying all the wages and running costs for a whole year, comes to $US300 million.

For the money spent on weapons last year, you could set up 3333 major hospitals around the world – one hospital for every 1½ million people. So in Australia alone, that's 12 new hospitals – each year.

So one year you could set up 3333 major hospitals, another year you could build 3333 large universities, another year you could build hundreds of thousands of community health centres, another you could build millions of baby health centres, and another, tens of millions of schools.

We would have a very different world if all the money that was spent on weapons in the last 10 years was spent on people.

bags, which would be puncture-proof, would remain inflated for as long as wanted. They should be available around 1996 or 1997.

The Pentagon sees *laser rifles* being used on a battlefield to temporarily blind the enemy soldiers. But if you move the beam of the laser rifle fairly slowly, it could play on somebody's eye for too long, and make them permanently blind. These disgusting weapons already exist, and at least one country will sell them right now.

In the USA, one-quarter of the police officers who die in the line of duty do so when their own *hand guns* are used against them. So one line of research at Sandia National Laboratory involves making hand guns keyed to the owner, so they can't be used by anyone else. It is relatively easy, using a sensor, to detect the size of the operator's hand. But hand size should be integrated with other readings unique to the licensed operator – such as voice, palm prints, or fingerprints.

Lawrence Livermore Laboratory has been working on 'Lifeguard'. It is a system which, in less than one second, finds the location of a hidden sniper. It could easily be linked to a system to automatically shoot back at the sniper.

The CIA has been testing *hallucinogenic drugs* as non-lethal weapons for the last quarter of a century, but the problem with various drugs is getting the dose right. It's all too easy to overdose someone and kill them.

And there's still the use of *foul-smelling gases* to drive people away.

One popular device to be used against **enemy weaponry** is the *very caustic chemical*, something like a super-concentrated dishwasher powder. These chemicals are more corrosive than the most powerful acids, and could eat through

Weapons Spending ②

the tracks on a tank. On the other hand, they would be nasty to human skin. The caustic chemicals do exist, but they're difficult to deliver to where you want them to go. One trouble is that they try to eat through the container that you hold them in!

Micro-organisms such as certain bacteria will eat virtually anything on the planet, including petrol, rubber and bitumen. But using bacteria and viruses as weapons is forbidden by the 1972 Biological Weapons Convention.

A *computer virus* could disable enemy computers and communications gear. One such virus was supposedly introduced into Iraq in the Gulf War – in a printer! One easy target is the banking business and finance of a foreign power – but such a weapon is also cheap for the enemy to make. Apparently the CIA has considered the use of such a weapon in the past, but has always ruled against its use, because it would be too easy for an enemy to use it in retaliation against the USA.

Microwave pulses, if they were intense enough, could melt delicate electronic components. Giant *electromagnetic pulses* could do the same thing. A pulse device has been tested once at the Los Alamos National Laboratory, but it's way too big to use on a battlefield.

A smaller application of a previous concept, an *electronic roadblock*, has already been invented to make car chases safer, by immobilising the suspect vehicle. The Anti-Vehicle Electronic Counter-Measure has been made by Creative Electronic Consultants of Sleepy Hollow, Illinois, USA. It looks like a large plastic pancake, about 1 metre across, resting on the road. As the offender's car passes over it, it gives off an electrical discharge that scrambles the car's electronics. It probably won't work on older cars that don't have computers, or on diesel cars.

During the Gulf War, a US Air Force missile showered millions of tiny *fibres of carbon* onto an electrical power plant in Baghdad. The carbon fibres were conductive, and they shorted out the power plant! Many industries, including a hospital, shut down. Indirectly, this 'non-lethal' weapon could have easily killed many civilians – but fewer than would have been killed by an air raid on the power plant.

Super-slippery fluids could be sprayed on to a runway so that planes couldn't land on it – these chemicals do exist, but there's not yet a really good way of delivering them that's any better than a modified crop duster plane. The same delivery problem holds for *combustion inhibitors*, which would stop the burning of petrol or diesel inside an engine. In practically all the recent wars, a combustion inhibitor could have nipped any conflict in the bud. Other non-lethal weapons to be used against enemy weaponry include *fast-hardening chemicals*, such as a stiff, expanding foam.

Unfortunately, as in most conflicts, it would be the civilians who stand to suffer the most from the use of non-lethal weaponry. For some reason, this whole field of non-lethal weaponry reminds me of the old saying of the anarchists: 'The state is an organisation which exists to imprison its subjects'.

References
New Scientist, No. 1903, 11 December 1993, 'War over weapons that can't kill' by Vincent Kiernan, pp.14–16.
Scientific American, April 1994, 'Bang! you're alive', pp.12, 13.
Aviation Week and Space Technology, 15 August 1994, 'Without even trying', p.13.
Popular Science, October 1994, 'Soft kill' by Robert Langreth, pp.66–69.
Wired, February 1995, 'Surrender or we'll slime you' by Mark Nollinger, pp.90–100.
Omni, February 1995, 'Future firearms' by Carol Silverman Saunders, p.31.

RUNNING ROACHES

HARDLY ANYBODY LOVES or cares about the cockroach – it's always had very bad press. In fact, most people don't even want to know about cockroaches. In the last few years, however, scientists have finally looked closely at how cockroaches run – and not only have they found that the cockroach is the world's fastest running insect, they're also using what they've learnt to make robots that will eventually work under the sea, and on other planets.

Now wheels are a pretty good way to get around. But so far, we've discovered only one animal that uses a wheel-like motion to get around. It's a prawn-like creature on the *flat* beaches of Panama. Wheels are OK on the flat, but when the going gets rough, legs are the way to go – and usually, the more the better. So the engineers who design robots to de-commission nuclear power plants, or crawl into volcanoes, or explore Mars usually copy insects. But while insects can move with great speed, style and agility, our multi-legged robots are very slow, quite clumsy, and not at all agile. How come? It's all to do with how animals get around.

It seems that when the robot designers copied insects, they didn't do a very good job. In fact, the robot scientists didn't *know* how insects walked, so they just copied how they *thought* insects walked.

Now people have always been curious how animals get around. In 1872, Leland Stanford, the founder of Stanford University who got fabulously wealthy from his railroads, got involved in a discussion as to whether a running horse ever

Largest Cockroach

According to the *Guinness Book of Records*, the largest cockroach in the world is the *Megaloblatta longipennis*, which comes from Columbia. One specimen has been measured at 97 millimetres long and 45 millimetres across!

had all four legs up at the same time. Apparently, he even bet $US25 000 on the outcome. But in 1872, nobody knew the answer. The only way to really find out was with photographs, so Stanford paid Eadweard Muybridge, the famous landscape photographer, to take some pictures.

It took Eadweard Muybridge about five years to get the answer. (He spent much of 1877 successfully defending himself from the charge of having killed his wife's lover!) First he built a runway for the horses to trot or gallop on. Then he set up a series of threads across the runway. Along one side of the runway, he set up a series of cameras that were triggered by the horse breaking these threads, as it rushed past. But the technology of 1872 had neither fast cameras nor fast film, so the first pictures were too dark and too blurry. However, he kept at it, and in 1877, he finally developed a photo that showed a horse with all four legs up in the air!

A horse has four legs to give it a rather secure footing, but we humans have only two legs. It has been said that we humans walk in a series of controlled falls, from which we're continually rescuing ourselves.

It turns out that when we humans want to move, we use two quite different styles – the 'pendulum' and the 'pogo stick'. We use the 'pendulum' for walking (see 'Walking Women'), and the 'pogo stick' for running.

A pendulum, on its way down, is driven by gravity. By the time it has reached the lowest point in its arc, it has turned its gravitational energy into kinetic energy. That kinetic energy pushes the pendulum up out of the gravity well, simultaneously converting kinetic energy into gravitational energy, until it briefly comes to a complete halt. At that stage, it has no more kinetic energy, but

Cockroach Stomp

Why are cockroaches so hard to stomp on? Because cockroaches can detect the air currents created by your moving foot or, if you're not particularly squeamish, your moving hand.

On the back end of the cockroach, two little organs which look like little sticks poke up into the air. These organs are called cerci (plural), and each cercus (singular) has 220 hairs. Each hair is very fragile and is easily moved by air currents. Depending on which way the air currents move, the little hairs on each cercus move, and this produces different electrical signals.

Rather than going to the brain, and then all the way back to the legs, these signals go directly to the back legs, which are very close to each cercus. These nerves are large and the distance is short, so the electrical information travels very rapidly – in something like 0.011 of a second. The cockie starts to actually move by 0.054 of a second.

This means that if you want to stomp a cockroach, you must do it from practically straight above. That way the wind current confuses the cockroach, which doesn't know which way to go.

It also helps to learn flamenco dancing. Dead cockroaches, however, impair the tone of the castanets.

Cockroach as Food

There is a balance in nature. While we humans have great difficulty in getting rid of cockroaches, the baby North American jewel wasp just eats them!

To keep her babies well fed, the caring mother North American jewel wasp lays each egg inside the body of a different cockroach. To make sure that the cockroach doesn't interfere with her plans, she permanently paralyses the cockroach before implanting her egg. The egg hatches, and then eats the cockroach from the inside, pupates, and finally comes out as a brand new jewel wasp.

This is an excellent way of providing food for baby. If she planted the egg inside a dead cockroach, it would decay and spoil, and baby would probably not survive. But her venom is very selective – it paralyses the leg muscles and the wing muscles, but leaves the heart muscles alone!

A Brief History of Insect Walkies

Galileo used a microscope as well as a telescope. He found that insects could walk upside-down because of little sucker pads on their feet.

Giovanni Borelli published a book called *De Motu Animalium* ('On The Movement Of Animals') in 1680. He rightly claimed that insects should move their six legs in a double-tripod gait. One tripod was the right front, right rear and left middle legs. The other tripod was the left front, left ➤

it does have its maximum possible amount of gravitational energy. Then it swings back down again, and the cycle continues. A pendulum is continually recycling gravitational energy into kinetic energy, and back again. When we humans walk, our body is a bit like an upside-down pendulum, and we're continually using gravitational energy to power us into our next step.

When we run, though, we begin behaving like a pogo stick, with our ligaments, tendons and muscles acting as springs which stretch and then compress. A speeding marathon runner will move their centre of gravity up and down by about 6 centimetres. As we run, energy is being stored in muscles, ligaments and tendons at the end of one step and is released at the beginning of the next step. The Achilles tendon stretches by about 5 per cent. Some of the energy is also stored in the arches of your feet as they continually flatten out and regain their shape – over and over again.

But everybody thought that animals with lots of legs (insects, crabs, centipedes and so on) moved along perfectly smoothly, with no bouncing up and down.

In the late 1980s, however, Robert Full, a neuro-physiologist who was then at Harvard but now runs the PolyPEDAL Laboratory at the University of California at Berkeley, began looking at how crabs run. (PEDAL is short for **P**erformance, **E**nergetics, and **D**ynamics of **A**nimal **L**ocomotion.) He used force plates. A force plate is a flat plate that – when you step on it – can measure up-and-down forces, and left-and-right forces, and forward-and-backward forces. When he analysed how the crabs were running, he realised they were like us humans. As they scuttled along, they were continually storing and releasing energy. At low speeds they behaved like a pendulum, but at higher speeds, they behaved like a pogo stick.

A crab is just one type of six-legged creature, though, so Robert Full worked with his student Michael Tu to set up a series of force plates to work with another six-legged creature, the cockroach. The cockroach is so small and light that everything had to be built much more delicately and precisely. It took them 12 months to set up the tiny force plates.

Once they started working with the cockies, though, they found a whole bunch of surprises. Before their research, everybody thought that all the cockies' legs had basically the same job. But Full and Tu found that the cockroach's rear legs were used to push the cockroach forward, while the short front legs were actually brakes which were used to slow the cockroach down. The middle legs could be used to either speed up or slow down the cockroach.

In the middle of all this serious laboratory work, they made it into the *Guinness Book of Records*! They measured the American cockroach (*Periplaneta americana*) running at 5 feet per second (about 5.4 kph), which makes it the fastest running insect on Earth. It can cover 50 times its own body length in one single second, which for a human works out to running at 320 kph.

But then they discovered another thing. Every now and then, the force plates would register absolutely no force. How could a six-legged creature have *all* of its six legs off the ground, even for a brief instant? To their surprise, the high speed video cameras showed that when the cockroach was trying to run as fast as it could, it would sprint on its two back legs! When we humans run, at some stage in each step, we

▷ rear and right middle legs. He said they should first move one tripod, and then the other tripod, and so on. At some speeds, some insects do move like this. He was right, but he had not done the experiment of actually looking at how they really moved. He had argued on theoretical grounds!

A Brief History of Cockroaches

Cockroaches are one of the great survivors. They have been around, essentially unchanged, for 340 million years. Back then, they made up about 40 per cent of the known insects, but today they are down to 1 per cent. Some scientists even say that cockroaches might have been the first flying animal.

They belong to an order of insects, Orthoptera, which contains some 23 000 different species. All insects in this order have straight front wings that protect the rear wings, and chewing mouthparts.

Cockroaches are long-legged, flat, oval insects. There are about 4000 different species worldwide, of which 450 are native to Australia. Less than 1 per cent of these species live in our houses, and are thought of as pests.

Dante II's Inferno

Dante I was a large robot that was supposed to explore places where it was too dangerous for humans to go. Its first mission was to explore Mt Erebus in Antarctica. However, on New Year's Day in 1993, it got less than 9 metres before its cable (providing both communications and power) kinked and broke!

Dante II was much improved. It had two sets of four legs, was about 3 metres long, weighed about 1 tonne, and cost $US1 700 000. Each set of four legs was attached to a moving plate. First one set of four legs moved forward and found a solid footing, then the second set moved, and so on. (All the legs were identical!) It could run on its own internal computers, making its own decisions as to where to go.

Its first mission was to explore a crater on the side of Mt Spurr, a 3374 metre-high volcano some 150 kilometres west of Anchorage, in Alaska. This crater, Crater Peak, some 800 metres across, was about 2300 metres above sea level. Mt Spurr had been quiet for 39 years, and then had erupted three times in 1993. Vulcanologists were interested in measuring levels of hydrogen sulphide and sulphur dioxide. In 1993 and 1994, eight geologists had died attempting such dangerous tasks – so it would be better for a machine to do the job.

Dante II had one video camera on each corner, plus another two close together and high up to give stereo vision. For its first mission, it was powered by a diesel generator just outside the lip of the crater. The 300 metre cable carried 1000 Volts AC, optic-fibre communications, and incorporated the incredibly strong synthetic fibre Kevlar. It sent pictures back through the cable to the satellite dish on the side of the volcano, up to a satellite, down to the scientists at Anchorage, and then out to the whole world via dedicated optic-fibre links and the Internet.

By the end of Day 2, it had gone about halfway down the wall of the 200 metre deep crater. On Day 3, after braving falling rocks as big as refrigerators and continual earthslips, it reached the bottom, and stayed there for a day. At the beginning of Day 4, a 30 centimetre rock smashed into one of Dante II's legs. By Day 6, it was halfway up the wall again – but navigation was difficult because the continual earthslips had changed the lie of the landscape since Dante II had gone down. Suddenly, the electrical power connection from the diesel generator to the cable failed. A helicopter rushed a team in to repair the fault.

On Day 7, only 120 metres from the rim, Dante II flipped itself over. On unstable ground, one leg had extended itself so far in an effort to find solid ground that it had tipped itself onto its back.

A helicopter tried to lift Dante II by its cable, but the cable broke, dumping it further down the hill. The second helicopter rescue attempt was successful, however.

Cockroaches and Pocahontas

The man who was saved by Pocohontas was the first man to use the word 'cockroach' in the English language!

The English word 'cockroach' comes from the Spanish word for cockroach – *cucaracha*. The first known use of the word in the English language is by the colonist Captain John Smith in 1624.

Fourteen years earlier, in December 1607, John Smith went up the Chicahominy River to try to obtain some corn from the Indians. According to his writings, he was captured by the Powhatan chief, Wahunsonacook, and was saved only because of the insistence of the chief's daughter, Pocohontas. She was 12 at the time.

In 1613, she was captured by the Virginia colonists and held for ransom. She adopted Christianity, changed her name to Rebecca, and on 5 April 1614, she married a Virginian widower, John Rolfe. She was 18 years old. She went to England, never seeing her father again. She had a son when in England, but died in a smallpox epidemic in 1617.

have both our feet in the air. The force plate was registering a zero force whenever the cockie was at that similar stage in its two-legged step, when both its legs were up in the air.

But why would the cockie run on two legs? The scientists think that this is because the front and middle legs of the cockroach are shorter than the rear legs. At the cockie's top speed, all six legs are already going as fast as they can – 27 strokes each second. The only way the cockie can go faster is to take the front and middle legs out of the circuit, so the longer back legs can take bigger strides.

But how could a cockroach run on just the two rear legs without falling flat on its face? The answer lies in aerodynamics. The body of a cockroach is flat on the bottom and curved on the top (just like an aeroplane wing), and this gives it some lift.

Who would have thought that cockroaches can sometimes be like us – two-legged?

Now if the engineers who have been building robots with legs a bit like insects had talked to the insect scientists, they'd be a lot better off. If they had bothered to check the excellent design of the insects *before* they tried to copy them, then maybe their slow and clumsy multi-legged robots would have been able to get off the ground, and go everywhere, man!

References

Exploring the Secrets of Nature, Reader's Digest Association Far East Limited, 1994, p.54.

Discover, September 1994, 'See how they run' by Carl Zimmer, pp.64–73.

New Scientist, No. 1946, 8 October 1994, 'The mathematical springs in the insect steps' by Jim Collins and Ian Stewart, pp.36–40.

New Scientist, No. 1951, 12 November 1994, 'Look to the insect...' by Kurt Kleiner, pp.27–29.

Popular Science, November 1994, 'Dante's inferno' by Judith Anne Gunther, pp.66–68, 96.